Study Guide

Chemistry

Principles, Patterns, and Applications
Averill • Eldredge

Karalyn Humphrey

Baylor University

PEARSON

Benjamin Cummings

San Francisco • Boston • New York
Cape Town • Hong Kong • London • Madrid • Mexico City
Montreal • Munich • Paris • Singapore • Sydney • Tokyo • Toronto

Publisher:	Jim Smith
Project Editor:	Kate Brayton
Editorial Assistant:	Kristin Rose Field
Managing Editor, Production:	Corinne Benson
Production Supervisor:	Jane Brundage
Senior Marketing Manager:	Scott Dustan
Marketing Development Editor:	Josh Frost
Manufacturing Buyer:	Pam Augspurger
Production Services:	Progressive Publishing Alternatives
Compositor:	Progressive Information Technologies
Cover Design:	Seventeenth Street Studios

Cover Photo Credit: B-Z reaction, Fritz Goro

ISBN 0-8053-3814-4

1 2 3 4 5 6 7 8—CRS—10 09 08 07 06
www.aw-bc.com

Contents

Introduction to Chemistry

Key Words

α particle	gas	noble gas
β particle	group	nonmetal
γ ray	halogen	nucleus
alkali metal	heterogeneous	observation
alkaline earth	homogeneous	period
atom	hypothesis	periodic table
atomic mass	intensive property	physical property
atomic mass unit	ion	proton
atomic number	isotope	qualitative observation
chemical property	law	quantitative observation
chemistry	liquid	radioactivity
compound	main group element	scientific method
density	mass	semimetal
electron	mass number	solid
element	matter	theory
essential element	metal	transition element
experiment	mixture	weight
extensive property	neutron	

By the end of this chapter, you should be able to:

- Define what chemistry is and give several examples of routine applications
- Describe the scientific method
- Classify matter
- Understand the history of chemistry as we've come to know it
- Understand basic radioactivity
- Understand atomic masses and isotopes
- Read a periodic table and understand its trends

Chapter Overview

Chemistry is an important subject matter, regardless of whether or not it's your major field of study. The science of chemistry influences many things in daily life and many of the current issues facing us as a society. Having a general feel for chemistry will help you understand much of how the world works, as well as many of the items you use on a daily basis.

1.1. Chemistry in the Modern World

Matter is anything that has mass and takes up space. *Chemistry* is defined as the study of matter and the changes that it undergoes. It is a very broad discipline that touches on many other fields, such as physics, biology, geology, forensics, engineering, and astronomy, just to name a few. Perhaps the most popular field of chemistry right now is forensics, which uses a wide variety of chemical methods to solve crimes. One of the most interesting recent topics in chemistry is the discovery that an asteroid impact may have caused the extinction of the dinosaurs 66 million years ago.

1.2. The Scientific Method

When approaching a question or problem, scientists follow a general method called the *scientific method*. This is a cyclical method that can be entered from any point and repeated as often as necessary. The method begins by making observations from which you form a hypothesis to try and explain what has been observed. Experiments are performed to test the hypothesis, and these experiments can cause the cycle to repeat with further observations and hypotheses. This method can lead to the formation of a law, which describes what happens, or to a theory, which attempts to describe why something happens.

1.3. A Description of Matter

There are several different ways to classify matter. One way is by its physical state of solid, liquid, or gas. Solids have a definite shape and volume; liquids have a changing shape and definite volume; and gases have both a changing shape and a changing volume. Matter can also be classified as a pure substance or a mixture. A pure substance can be an element, which cannot be broken down further by chemical means, or a compound, which can be broken down further by chemical means. A mixture can be homogeneous, having the same composition throughout, or heterogeneous, not having the same composition throughout. Two types of properties are used to describe matter: physical and chemical. Physical properties can be observed without changing the chemical identity of the material. Chemical properties can be observed only by changing the chemical identity of the substance. Physical properties can further be broken down into intensive and extensive properties. Intensive properties do not depend on sample size, while extensive properties do.

1.4. A Brief History of Chemistry

The first idea of the atom came about in ancient Greece in the fourth century B.C. Up until that time, it was believed that matter consisted of the four basic elements of fire, water, earth, and air. For about the next 2,000 years, advances in chemistry came from a group known as the *alchemists,* who were part scientist and part philosopher and who attempted to turn lesser metals into gold. What we know as modern chemistry began to develop in the 16th and 17th centuries through advances in metallurgy and the discovery of elements. Antoine Lavoisier wrote the first chemistry textbook and is credited with being the father of modern chemistry through his law of the conservation of mass. Proust and Dalton developed further fundamental chemical ideas that ultimately led to Dalton's atomic theory of matter. Of particular importance was the work of Amadeo Avogadro, who formulated the idea that equal volumes of a substance contained equal numbers of particles.

1.5. The Atom

An *atom* is defined as the smallest particle of an element that displays the properties of that element. Atoms are composed of negatively charged electrons that orbit around a nucleus of positively charged protons and electrically neutral neutrons. Radioactivity was discovered through the work of Henri Bequerel and Marie and Pierre Curie, who found that certain elements emitted a new form of energy. Radiation can take the form of α particles, which are helium nuclei; β particles, which are electrons traveling at a high rate of speed; and γ rays, which are energy rays similar to X-rays but with a higher energy.

1.6. Isotopes and Atomic Masses

When you look at a periodic table, you see that each element is identified by an atomic number. This is the number of protons contained in the nucleus of that element's atom. A neutral atom of an element will have the same number of electrons as protons. An atom with a different number of electrons is known as an *ion,* with a positive ion having fewer electrons than protons and a negative ion having more electrons than protons. If the number of neutrons is changed, this makes an isotope of that element. All isotopes of an element have the same atomic number but a different mass number, which is the sum of protons and neutrons. The weighted average of all the isotopes of an element is given as the atomic mass of the element. This mass is expressed in atomic mass units (amu), which is defined as one-twelfth of the mass of one atom of carbon-12.

1.7. Introduction to the Periodic Table

Perhaps the most useful tool available to chemistry students is the periodic table, which is an arrangement of all the currently known elements in order of increasing atomic number. The periodic table is composed of eighteen vertical columns called *groups,* whose elements exhibit similar chemistry, and seven horizontal rows called *periods.* Four of the most commonly used groups are the alkali metals (Group 1), the alkaline earths (Group 2), the halogens (Group 17), and the noble gases (Group 18). The elements can also be grouped into the mail group elements, Groups 1, 2, 13–18; the transition elements, Groups 3–12; the lanthanides; and the actinides. The elements can be classified as metals, nonmetals, or semimetals. Metals and nonmetals exhibit distinctly different properties and chemistry, and semimetals exhibit properties and chemistry that lie between those of metals and nonmetals.

1.8. Essential Elements

There are about 115 known elements. Of these, approximately 19 are currently believed to be essential for humans, without which normal biological function and growth cannot occur. We are composed primarily of carbon, hydrogen, oxygen, nitrogen, and sulfur. In addition, we need such elements as calcium, magnesium, sodium, potassium, chlorine, and phosphorous. The rest of the essential elements are present in trace amounts. The dietary intake of these elements varies greatly from one to another, and all of these elements range from deficient to optimum to toxic with increasing intake.

Self-Test

1. Chemistry is primarily concerned with the study of _____
   A. energy.
   B. light and its properties.
   C. matter and the changes it undergoes.
   D. imbalances in the human brain.

2. One of the most rapidly growing areas of chemistry is _____
   A. astrochemistry.
   B. forensics.
   C. geochemistry.
   D. chemical engineering.

3. The process of making observations, forming hypotheses, and performing experiments is known as the _____
   A. qualitative observation.
   B. law of definite proportions.
   C. scientific method.
   D. theory.

4. A statement that attempts to explain why something happens is a(n) _____
   A. hypothesis.
   B. theory.
   C. law.
   D. analysis.

5. Which of the following is not a pure substance?
   A. Sodium chloride
   B. Bromine
   C. Oxygen
   D. Tap water

6. Which of the following is not a physical property?
   A. Color
   B. Smell
   C. Flammability
   D. Volume

7. Which of the following is not an intensive property?
   A. Volume
   B. Melting point
   C. Boiling point
   D. Conductivity

8. Who is considered to be the father of modern chemistry?
   A. Robert Boyle
   B. Joseph Priestly
   C. Antoine Lavoisier
   D. John Dalton

9. Who proposed that equal volumes of different gases contain equal numbers of gas particles?
   A. Robert Boyle
   B. John Dalton
   C. Joseph Gay-Lussac
   D. Amadeo Avogadro

10. Particles of an atom that have a negative charge are the _____
    A. protons.
    B. neutrons.
    C. electrons.
    D. γ rays.

11. The most penetrating form of radiation is the _____
    A. α particle.
    B. β particle.
    C. γ ray.
    D. nucleus.

12. In the current model of the atom, electrons occupy _____
    A. a sphere of uniform positive charge.
    B. fixed circular orbits.
    C. orbitals determined by their energy.
    D. orbitals determined by the size of the nucleus.

13. Atoms that have the name number of protons but different numbers of neutrons are called _____

    A. ions.

    B. isotopes.

    C. transition elements.

    D. semimetals.

14. What is the weighted average of the masses of the isotopes?

    A. The atomic mass unit

    B. The atomic number

    C. The atomic mass

    D. The mass number

15. Which of the following is not an alkali metal?

    A. Lithium

    B. Potassium

    C. Cesium

    D. Magnesium

16. Which of the following is not a halogen?

    A. Bromine

    B. Chlorine

    C. Argon

    D. Fluorine

17. The vertical columns in the periodic table are called _____

    A. periods.

    B. groups.

    C. transition elements.

    D. lanthanides.

18. The elements in the upper right-hand corner of the periodic table are _____

    A. metals.

    B. nonmetals.

    C. semimetals.

    D. actinides.

19. Which of the following is not considered an essential element?

    A. Oxygen

    B. Magnesium

    C. Arsenic

    D. Osmium

20. Approximately how many of the known elements are considered essential?
    A. 11
    B. 19
    C. 22
    D. 25

Answers: 1. C; 2. B; 3. C; 4. B; 5. D; 6. C; 7. A; 8. C; 9. D; 10. C; 11. C; 12. C; 13. B; 14. C; 15. D; 16. C; 17. B; 18. B; 19. D; 20. B

CHAPTER 2

Molecules, Ions, and Chemical Formulas

Key Words

acid	chemical nomenclature	intramolecular
alcohol	condensed structural	ion
aliphatic	formula	ionic compound
alkane	covalent bond	molecular formula
alkene	covalent compound	molecule
alkyl group	cracking	monatomic ion
alkyne	cyclic hydrocarbon	octane rating
amine	double bond	organic compound
anion	electrostatic attraction	oxoacid
aromatic (arene)	electrostatic interaction	polyatomic ion
aryl group	electrostatic repulsion	reforming
base	empirical formula	saturated
binary covalent compound	formula unit	single bond
binary ionic compound	hydrate	structural formula
carboxylic acid	hydrocarbon	triple bond
cation	inorganic compound	unsaturated
chemical bond	intermolecular	

By the end of this chapter, you should be able to:

- Identify ionic and covalent bonding
- Understand the forces holding compounds together
- Describe the composition of a chemical compound
- Understand the difference between structural, condensed, and empirical formulas
- Write the empirical formula of a compound
- Understand the difference between organic and inorganic compounds
- Name several broad groups of organic compounds
- Name ionic compounds
- Name simple covalent compounds

- Identify and name common acids and bases
- Discuss such industrially important chemicals as petroleum and sulfuric acid

Chapter Overview

The previous chapter introduced a number of the fundamental ideas of chemistry, particularly the concepts of atoms and elements. This chapter builds upon those concepts by introducing you to chemical compounds, mixtures of atoms, and elements. There are 115 currently known elements, but there are millions of currently known compounds. Being able to recognize and name different kinds of compounds is essential as you continue your study of chemistry.

2.1. Chemical Compounds

The atoms that make up chemical compounds are held together by forces known as *electrostatic interactions,* which can be either attractive or repulsive. These interactions make up the chemical bond. Compounds that are made up of positively and negatively charged ions are called *ionic compounds.* Compounds that are made up of atoms that share electrons between each other are called *covalent compounds,* and they consist of molecules. Molecules are held together by electrostatic attractions between the positive nuclei and the negative electrons. Chemists express what kinds of atoms are present in a compound, and how many, by writing a *molecular formula.* Organic compounds are made up of primarily hydrogen and covalently bound carbon. Inorganic compounds are primarily made up of other elements. There are several different kinds of formulas available to chemists. Structural formulas show the composition and 2-D bonding of the molecule. Three-dimensional formulas show the shape of a molecule in space, and also sometimes show the relative sizes of the component atoms. Perhaps the most frequently used formula is the condensed formula, which lists the atoms bonded to one another. Covalent bonds can be either single, sharing one pair of electrons; double, sharing two pair of electrons; or triple, sharing three pairs of electrons. Ions can have a positive charge, called a *cation,* or a negative charge, called an *anion.*

2.2. Chemical Formulas

The chemical formula of a molecule describes the atoms that are in a molecule. The empirical formula gives the lowest whole numbers of elements in a compound—essentially, the ratio of one element to another. The formula unit is the absolute grouping of elements represented by the empirical formula. Empirical formulas can be of particular use when dealing with ionic compounds. There are ions, known as *polyatomic ions,* that are groups of atoms that bear electrical charge, though the atoms themselves are covalently bonded to one another. Some ionic compounds occur as hydrates, which are compounds containing a certain number of water molecules, called *waters of hydration.* These waters can be easily gotten rid of, revealing the pure compound.

2.3. Naming Ionic Compounds

Many ionic compounds have both a common and a systematic name. While the common name may come from any number of sources, the systematic name follows a definite pattern from compound to compound. We can write the chemical formula from the systematic name, and vice versa.

To name an ionic compound, use the following steps:

1. Place the ions with cations first, followed by anions.
2. Name the cation.
    a. If a metal forms only one kind of ion, simply name it.
    b. If a metal forms more than one kind, indicate it by a roman numeral in parentheses.
    c. Polyatomic cations have names that must be memorized.
3. Name the anion.
    a. Monatomic anions end in –*ide*.
    b. Polyatomic anions have names that must be memorized.
4. Write the name of the compound as the name of the cation followed by the name of the anion.

2.4. Naming Covalent Compounds

Like ionic compounds, covalent compounds may possess both a common and a systematic name. The procedure for naming a covalent compound is very similar to that for naming an ionic compound.

To name a covalent compound, use the following steps:

1. Place the elements in their proper order.
    a. Element farthest to the left is named first.
    b. Second element is named as if it were ionic, ending in –*ide*.
2. Identify the number of each type of atom present, using appropriate prefixes.
3. Write the name of the compound.
    a. Certain compounds are exceptions and are always named by their common names.

Organic compounds are important to both industry and biological functions. The simplest of the organics are the hydrocarbons, composed of carbon and hydrogen. Aliphatic hydrocarbons are straight-chained molecules that can have single bonds (alkanes), double bonds (alkenes), or triple bonds (alkynes). Hydrocarbons can also be cyclic, meaning that the chain joins to form a ring. Aromatic hydrocarbons are a special class that has rings derived from the structure of benzene. Alcohols are an important organic functional group, often used in solvents and fuels.

2.5. Acids and Bases

In this chapter, we define acids as substances that dissociate to form an anion and H^+ ions in water. Bases are defined as substances that dissociate to form a cation and the OH^- anion in water. Neutral solutions are solutions that are neither acids nor bases. Organic acids are carboxylic acids.

Names for acids come from the names of their anions:

1. If the name of the anion ends in –*ate,* the acid name ends in –*ic.*
2. If the name of the anion ends in –*ite,* the acid name ends in –*ous.*

Most bases are composed of metal cations attached to hydroxide ions. They are named for the metal cation plus *hydroxide.* The common exception to this rule is ammonia, which is a weak base. Organic bases are amines.

2.6. Industrially Important Chemicals

The chemical industry produces hundreds of billions of pounds of chemicals annually for use in food production, plastics and other materials, and fuels. Petroleum refining is a particularly long and complex procedure that converts a mix of hydrocarbons into products suitable for use as fuels and in industry. In the process of cracking, heavier hydrocarbons are heated to a high temperature and broken down into lighter hydrocarbons. The process of reforming works to convert straight-chain alkanes to branched-chain molecules or to mixtures of aromatic molecules. The mixtures obtained through these processes are separated out through fractional distillation. The lighter products are used as fuels and oils, while the heavier products are used as lubricants and asphalt. The major use of petroleum products is in the production of gasoline. The quality of gasoline is indicated by the octane rating, which is the measure of the fuel's ability to burn cleanly and smoothly in a combustion engine. A gasoline's octane rating depends upon which compounds are present in the fuel and their relative abundance.

Sulfuric acid is the top industrial chemical produced in the world. Its primary use is in the production of fertilizer. The common process for obtaining sulfur from the earth is known as the *Frasch process,* in which 160°C high-pressure water forces sulfur out of porous limestone and up to the surface in molten and nearly pure form. This sulfur is then combusted to form sulfur dioxide. Sulfur dioxide is reacted with oxygen and vanadium (V) oxide to produce sulfuric acid.

Self-Test

1. When one or more pairs of electrons are shared between bonded atoms, this is known as a(n) _____
    A. ionic bond.
    B. covalent bond.
    C. electrostatic attraction.
    D. electrostatic repulsion.

2. Compounds that contain predominantly carbon and hydrogen are known as _____
    A. organic compounds.
    B. inorganic compounds.
    C. ionic compounds.
    D. covalent compounds.

3. Which of the following is not a way of representing a molecular structure?
   A. Condensed formula
   B. Structural formula
   C. Space-filling formula
   D. Dalton formula

4. The electrostatic energy of a molecule is dependent upon all of the following except _____
   A. magnitude of the particle's charge.
   B. sign of the particle's charge.
   C. arrangement of ions in space.
   D. internuclear distance.

5. What is the best empirical formula for magnesium oxide?
   A. MgO
   B. MgO_2
   C. Mg_2O
   D. Mg_2O_2

6. What is the correct formula for a compound of aluminum and oxygen?
   A. AlO
   B. Al_3O
   C. AlO_3
   D. Al_2O_3

7. What is the correct formula for the chlorate ion?
   A. ClO_4^-
   B. ClO_3^-
   C. ClO_2^-
   D. ClO^-

8. What is the correct formula for cupric sulfate?
   A. Cu_2SO_4
   B. $CuSO_4$
   C. Cu_2SO_3
   D. $CuSO_3$

9. What is the systematic name for $FeCl_3$?
   A. Iron (II) chloride
   B. Iron (III) chloride
   C. Ferrous chloride
   D. Ferric chloride

10. What is the correct formula for Lithium perchlorate?
    A. $LiClO_4$
    B. $LiClO_3$
    C. $LiClO_2$
    D. $LiClO$

11. Which of the following is a not a class of straight-chained hydrocarbons?
    A. Alkanes
    B. Alkenes
    C. Alkynes
    D. Aromatics

12. What is the correct name for the hydrocarbon with the formula $CH_3(CH_2)_3CH_3$?
    A. Propane
    B. Butane
    C. Pentane
    D. Isobutane

13. What is the correct formula for 3-heptene?
    A. $CH_3CH{=}CHCH_2CH_2CH_2CH_3$
    B. $CH_3CH_2CH{=}CHCH_2CH_2CH_3$
    C. $CH_3CH_2CH_2(CH_3)CH_2CH_2CH_3$
    D. $CH_3CH_2CH_2{=}CH_2CH_2CH_2CH_3$

14. What is the correct formula for dinitrogen tetroxide?
    A. NO_2
    B. N_2O
    C. N_2O_3
    D. N_2O_4

15. What is the correct name for $Ca(NO_2)_2$?
    A. Calcium nitrite
    B. Copper nitrate
    C. Calcium nitrate
    D. Nitrous calcite

16. An acid in which the H^+ is attached to an oxygen atom of a polyatomic anion is known as a(n) _____
    A. *–ic* acid.
    B. *–ous* acid.
    C. amine.
    D. oxoacid.

17. What is the correct name for $HClO_3$?
    A. Perchloric acid
    B. Chloric acid
    C. Chlorous acid
    D. Hypochlorous acid

18. What is the correct formula for calcium hydroxide?
    A. $Ca(OH)_2$
    B. $CaOH_2$
    C. $CaOH$
    D. Ca_2OH

19. Which of the following is not one of the steps of petroleum refining?
    A. Cracking
    B. Reforming
    C. The contact process
    D. Fractional distillation

20. The process by which hot water is used to force nearly pure sulfur from rock is the _____
    A. lead chamber process.
    B. Frasch process.
    C. contact process.
    D. oleum process.

Answers: 1. B; 2. A; 3. D; 4. C; 5. A; 6. D; 7. B; 8. B; 9. B; 10. A; 11. D; 12. C; 13. B; 14. D; 15. C; 16. D; 17. B; 18. A; 19. C; 20. B

Chemical Reactions

Key Words

acid–base reaction
actual yield
Avogadro's number
catalysis
catalyst
chemical equation
chemically equivalent
chemical reaction
cleavage reaction
coefficient
combustion analysis
combustion reaction
condensation reaction
enzyme
excess

exchange reaction
formula mass
heterogeneous catalyst
homogeneous catalyst
limiting reactant
molar mass
mole
molecular mass
mole ratio
oxidant
oxidation
oxidation state
ozone
ozone layer
percent composition

percent yield
product
reactant
redox reaction
reductant
reduction
stoichiometric quantity
stoichiometry
stratosphere
theoretical yield
troposphere
ultraviolet light
visible light

By the end of this chapter, you should be able to:

- Calculate the formula mass of an ionic compound
- Calculate the molecular mass of a covalent compound
- Calculate the number of molecules, formula units, or atoms in a sample
- Determine the empirical formula of a compound
- Determine the molecular formula of a compound from its empirical formula
- Write and balance chemical equations
- Name and recognize the fundamental types of chemical reactions
- Calculate the quantities of compounds produced or consumed in chemical reactions

Chapter Overview

The last chapter introduced concepts of ionic and molecular bonds and chemical compounds. This chapter follows that by introducing the concept of chemical reactions, the ways in which chemical compounds are formed and react with one another. Chemical reactions are vital and widespread in the world around us, and an understanding of them will go a long way toward comprehending many processes that you observe in your daily life.

3.1. The Mole and Molar Masses

There are several ways of describing the mass of a compound. The molecular mass is found by adding the average masses of the atoms in one molecule of that compound. For ionic compounds, the formula mass is the sum of the atomic masses of all the elements in the empirical formula, with each multiplied by its subscript. The units of both these expressions are the same—atomic mass units or *amus*. The mole is an arbitrary quantity established to measure the number of atoms, molecules, or formula units in a given mass of a substance. Also referred to as *Avogadro's number*, the mole is defined as the number of atoms of carbon-12 contained in 12 grams and is equal to 6.022×10^{23} atoms. From this definition comes the concept of the molar mass, which is the mass of one mole of a substance. The units for the molar mass are grams per mole.

3.2. Determining Empirical and Molecular Formulas

Although there are some exceptions to this rule, generally speaking a chemical compound always contains the same proportion of elements by mass. This rule allows us to calculate a percent composition, which is the percent of each element present in a pure substance. From this percent composition, we can determine the empirical formula, which gives the relative number of atoms in a substance in the smallest possible ratio. The most common way to determine the percent composition and empirical formula is by combustion analysis, where a sample is burned in oxygen and the resulting gases are collected and measured.

3.3. Chemical Equations

Chemical reactions are described by chemical equations, which are like stories telling the identities and quantities of substances involved in a reaction. A chemical equation reads like a sentence and shows the starting materials, or reactants, on the left and the resulting materials, or products, on the right, with the two sides separated by an arrow that indicates the transformation from reactants to products. A balanced chemical equation has the same number of atoms and the same total charge equal, or balanced, on both sides. The equation is balanced through the number of atoms, molecules, or formula units, known as the *coefficient* of that species. The mole ratio of two substances in a chemical reaction is the ratio of their coefficients in the balanced chemical equation.

In order to balance a chemical equation, use the following steps:

1. Identify the most complex substance.
2. Adjust the coefficient. Try to begin with an element that appears in only one reactant and one product, and work through the equation from there.

3. Balance polyatomics as a unit.

4. Balance the remaining atoms.

5. Check that both sides of the equation are now balanced.

3.4. Mass Relationships in Chemical Equations

Stoichiometry is the term describing the quantitative relationship between the masses, numbers of moles, and numbers of particles involved and formed in a chemical reaction. The stoichiometric quantity is the amount of product or reactant specified by coefficients in the balanced chemical equation. It is more practical to work with masses than moles, and stoichiometry allows us to move smoothly from one to the other.

To convert between masses of reactants and products, follow these steps:

1. Convert the mass of one substance (A) to the number of moles.

2. Find how many moles of another substance (B) comes from A using the mole-to-mole ratio (ratio of coefficients).

3. Convert the number of moles of B to the mass of B.

The reactant that limits the amount of product that can be made is known as the *limiting reactant*. Any reactant that is not completely consumed in the reaction is said to be present in excess.

To identify and work with limiting reactants, follow these steps:

1. Determine the number of moles of each reactant.

2. Compare the mole ratio of the reactants with the ratio in the balanced equation. From this, you can determine which reactant is limiting.

3. Calculate the moles of product that can be obtained from the limiting reactant.

4. Convert the moles of product to the mass.

The theoretical yield of a reaction is the maximum amount of product(s) that can be generated by a given amount of reactants. The actual yield is what you actually get in the reaction, and it cannot exceed the theoretical yield. The percent yield is calculated as the ratio of the actual yield to theoretical yield, multiplied by 100 to convert it to a percent; it can never be larger than 100%.

3.5. Classifying Chemical Reactions

There are four basic types of chemical reactions: acid–base, exchange, condensation or cleavage, and oxidation-reduction (redox). An atom's oxidation state is the charge that it would have if all its bonding electrons were transferred completely to the atom with the greater attraction for electrons. In a redox reaction, one compound must lose electrons, while another one must gain electrons. Oxidation is the loss of electrons, resulting in an increase in oxidation state. Reduction is the gaining of electrons, resulting in a decrease in oxidation state. Oxidants are those compounds that can accept electrons and be reduced in the reaction. Reductants are those compounds that can donate electrons and be oxidized during the reaction. A combustion reaction is essentially a rapid redox reaction in which the oxidant is the oxygen molecule. The rate of many chemical reactions can be increased through the use of a catalyst, which is a substance that participates in a reaction without being consumed. *Catalysis* is the process of accelerating a reaction through a catalyst. A homogeneous catalyst is one that is uniformly distributed through the reactant to form a solution. A heterogeneous catalyst is one that is in a different physical state than the reactants.

3.6. Chemical Reactions in the Atmosphere

The earth's atmosphere consists of several layers. The range of radiation produced by the sun includes visible light and ultraviolet light. In the stratosphere, ultraviolet light reacts with diatomic oxygen to produce atomic oxygen. This then reacts with a further oxygen molecule to produce ozone (O_3). This reaction leads to the formation of an ozone layer within the stratosphere. The absorption of ultraviolet light that is needed for this reaction helps to keep the light from penetrating through to the earth and causing damage. It has been recently discovered that organic compounds containing chlorine and fluorine, known as *chlorofluorocarbons* or *CFCs*, are volatile and reach the stratosphere to react with ultraviolet light themselves to produce chlorine compounds that convert the ozone back to oxygen molecules, thus reducing the ozone layer and allowing more ultraviolet light to reach the earth's surface. CFCs have been or are being phased out in many parts of the world and are being replaced with hydrofluorocarbons, which break down before reaching the stratosphere. It is hoped that this move will minimize further damage to the ozone layer.

Self-Test

1. Calculate the molecular mass of pentanol, $CH_3(CH_2)_3CH_2OH$.

   A. 63 amu

   B. 88 amu

   C. 63 g/mol

   D. 88 g/mol

2. Calculate the formula mass of $Be(NO_3)_2$.

   A. 117 amu

   B. 117 g/mol

   C. 133 amu

   D. 133 g/mol

3. How much do Avogadro's number of sulfur atoms weigh?

   A. 16 g

   B. 32 g

   C. 16 amu

   D. 6.02×10^{23} amu

4. What is the molar mass of one mole of nitrogen gas?

   A. 14 g

   B. 28 g

   C. 14 amu

   D. 28 amu

5. What is the percent carbon in ethanol, C_2H_5OH?

    A. 24%

    B. 26%

    C. 52%

    D. 54%

6. Combustion analysis is useful for compounds containing all of the following except _____

    A. carbon.

    B. nitrogen.

    C. sulfur.

    D. phosphorous.

7. Balance the following net ionic equation:

$$HCO_3^- + H_3O^+ \rightarrow \underline{} H_2O + \underline{} H_2CO_3$$

    A. 1,2

    B. 2,1

    C. 2,2

    D. 1,1

8. What is the sum of the coefficients in a properly balanced equation describing the combustion of ethanol to produce carbon dioxide and water?

    A. 9

    B. 8

    C. 7

    D. 6

9. What is the coefficient for water in the following reaction?

$$(NH_4)Cr_2O_7 \rightarrow Cr_2O_3 + N_2 + H_2O$$

    A. 2

    B. 3

    C. 4

    D. 5

10. The decomposition of potassium chlorate produces potassium chloride and oxygen. What is the coefficient for oxygen in the balanced equation?

$$KClO_3 \rightarrow KCl + O_2$$

    A. 2

    B. 3

    C. 4

    D. 6

11. How many moles of oxygen can be produced from one mole of $KClO_3$?

    A. 1

    B. 1.5

    C. 2

    D. 3

12. How many grams of oxygen can be produced from 100 g of $KClO_3$?

    A. 28

    B. 39

    C. 53

    D. 84

13. How many grams of CO_2 are produced from the complete combustion 100 g of propane?

    A. 100

    B. 200

    C. 300

    D. 400

14. How many grams of water are produced from the complete combustion of 100 g of propane?

    A. 327

    B. 222

    C. 164

    D. 123

15. Which of the following is not one of the major types of chemical reactions?

    A. Oxidation-reduction

    B. Acid–base

    C. Exchange

    D. Combustion

16. The oxidation number of P in $H_2PO_4^{-}$ is _____

    A. +5.

    B. +6.

    C. +7.

    D. +8.

17. Which of the following is not a form of a catalyst?

    A. Homogeneous

    B. Heterogeneous

    C. Reducing

    D. Enzyme

18. The formation of water from hydrogen gas and oxygen gas is what type of reaction?

    A. Acid–base

    B. Exchange

    C. Cleavage

    D. Condensation

19. Which element is most abundant in the earth's atmosphere?

    A. Hydrogen

    B. Oxygen

    C. Carbon dioxide

    D. Nitrogen

20. The element most likely related to the depletion of the ozone layer is _____

    A. fluorine.

    B. chlorine.

    C. bromine.

    D. argon.

Answers: 1. B; 2. C; 3. A; 4. B; 5. C; 6. D; 7. D; 8. A; 9. C; 10. B; 11. B; 12. B; 13. C; 14. C; 15. D; 16. A; 17. C; 18. D; 19. D; 20. B

CHAPTER 4

Reactions in Aqueous Solution

Key Words

acid rain
active metals
activity series
amphoteric
aqueous solution
Arrhenius acid
Arrhenius base
Brønsted–Lowry acid
Brønsted–Lowry base
complete ionic equation
concentration
conjugate acid
conjugate base
diprotic
electrical potential
electrolyte
endpoint
equilibrium
equivalence point
hydrated ions

hydronium ion
indicators
inert metals
insoluble
molarity
monoprotic
net ionic equation
neutralization reaction
neutral solution
nonaqueous solution
nonelectrolyte
overall equation
oxidation state method
parts per billion
parts per million
pH scale
polar bond
polyprotic
precipitate
precipitation reaction

quantitative analysis
salt
single-displacement
  reaction
soluble
solutes
solution
solvent
spectator ions
standard solution
strong acid
strong base
strong electrolyte
titrant
titration
triprotic
weak acid
weak base
weak electrolyte

By the end of this chapter, you should be able to:

- Understand solutions and how they form
- Calculate the concentrations of solutions
- Balance chemical equations of reactions in solution
- Understand and work with the stoichiometry of reactions in solution
- Understand the properties of and define acids and bases
- Understand precipitation, acid–base, and redox reactions in solution
- Understand the use of titration reactions in quantitative analysis

Chapter Overview

In the previous chapter, we began a general discussion of chemical reactions. The vast majority of chemical reactions occur in solution. We will therefore continue our discussion by considering solutions and the types of chemistry that can occur in solution.

4.1. Aqueous Solutions

A *solution* is defined as a homogeneous mixture of two or more substances. The solute, which is present in the smaller amount, is dispersed among the solvent, which is present in the greater amount. The use of water as the solvent defines whether or not the solution is aqueous or nonaqueous. The shape of the water molecule allows for an uneven charge distribution, with the oxygen carrying a partial negative charge while the hydrogen carries a partial positive charge, resulting in a polar molecule. A general rule of chemistry is that "like dissolves like," so polar substances and ionic compounds are most soluble in polar solvents like water. Substances dissolved in water are categorized by their ability to conduct electricity. A strong electrolyte dissociates completely into ions and produces a solution that strongly conducts electricity. A weak electrolyte does not dissociate completely, and the resulting solution is not a good conductor of electricity. A nonelectrolyte produces a solution that does not conduct at all.

4.2. Solution Concentrations

The amount of solute present in a solution is given as the solution's concentration. The most common expression of concentration is *molarity,* which is the number of moles per liter of solution. Two ways of expressing concentrations in very dilute solutions are parts per million (ppm) and parts per billion (ppb). A solution of a given molarity would be prepared by dissolving a known mass of the solute in a solvent and then diluting the solution to the appropriate final volume. This then could also serve as a stock solution, from which given amounts would be drawn and diluted further as needed.

4.3. Stoichiometry of Reactions in Solution

The stoichiometry involved in reactions occurring in solutions is fundamentally the same as what we studied in the previous chapter. The main difference is that, instead of working with masses of solids, we are now working with volumes of liquids. The molarity allows us to know how many moles are in a given volume; from there, we can apply the basic rules of stoichiometry that we've already learned (i.e., that the coefficients in the balanced chemical equation tell us the relationship between the reactants and the products).

4.4. Ionic Equations

There are three basic ways of writing the chemical equation of a reaction that occurs in solution. The *overall equation* shows all the products and reactants as their complete, neutral compounds. The *complete ionic equation* shows all of the reactants and products in the forms in which they actually exist in the solution, demonstrating which ions and molecules are hydrated and which are present in other forms and phases. The complete ionic equation will contain some spectator ions, which do not take part in the actual reaction. These are left out of the *net ionic equation,* which shows only those species that actually take part in the reaction. The three most common types of solution reactions are oxidation-reduction, acid–base, and precipitation.

4.5. Precipitation Reactions

In a precipitation reaction, the mixture of two solutions will form an insoluble solid known as a *precipitate.* The product of this type of reaction can be predicted if you know the solubilities of the species present and can determine if any combination of them will produce an insoluble salt.

4.6. Acid–Base Reactions

There are several ways to define acids and bases. Arrhenius defined an acid as a substance that dissolved to produce H^+ ions in water, and defined a base as a substance that dissolved to produce OH^- ions in water. Brønsted and Lowry defined an acid as a proton donor, and defined a base as a proton acceptor. It is the Brønsted–Lowry definition that is currently used. Monoprotic acids donate one proton, diprotic acids donate two protons, and triprotic acids donate three protons. Most generally, acids are classified as being either monoprotic or polyprotic. Acids and bases are also classified as being either strong or weak. A strong acid will dissociate completely in water to form H^+ ions, while weak acids will only partially (5% or less) dissociate in water. A strong base will dissociate completely in water to form OH^- ions, while weak bases will only partially do so. A strong acid and a strong base will react together in a double-displacement reaction known as a *neutralization reaction,* yielding the products of water and a salt. How acidic or basic a solution is can be described by using the pH scale. The pH is defined as the negative logarithm of the concentration of hydrogen ions. The pH for an acidic solution will be in the range of 0 to just below 7. A neutral solution will have a pH of 7. A basic solution will have a pH ranging from just over 7 to 14. The pH of a solution can be measured with an indicator, which is generally an organic substance that changes color in response to pH.

4.7. The Chemistry of Acid Rain

Normal, natural rain is slightly acidic, having a pH of 5.6 due to the presence of dissolved carbon dioxide. Rain is considered to be acid rain when the pH dips below 5.6; rain as low as pH 2 has been reported in the United States alone. The combustion of fossil fuels generates the vast majority of nitrogen oxides and sulfur dioxide that leads to acid rain, though a certain amount can also come from such natural phenomena as volcanoes, forest fires, and microbial decay. Over time, these oxides react with oxygen and water to generate nitric and sulfuric acid, which then rains back down to the earth and can cause such damage as eroding statues and structures and lowering stream and lake pH to the point of killing fish.

4.8. Oxidation-Reduction Reactions in Solution

In the previous chapter, we defined an oxidation-reduction reaction as one in which electrons are transferred from one compound or atom to another. Redox reactions that occur in solution are more complex than our previous examples. They can be balanced through a method known as the *oxidation state method,* which is given here.

Balancing a redox reaction by the oxidation state method:

1. Write the unbalanced equation for the reaction, showing the reactants and products.

2. Assign oxidation states to all atoms in the reactants and products and determine which atoms change oxidation state.

3. Write separate equations for oxidation and reduction, showing the atom(s) that is (are) oxidized and reduced, plus the number of electrons accepted or donated.

4. Multiply the oxidation and reduction equations by appropriate coefficients so that both contain the same number of electrons.

5. Write the equations showing the actual chemical forms of the reactants and products, adjusting the coefficients as necessary.

6. Add the two equations and cancel the electrons.

7. Balance the charge by adding H^+ or OH^- ions as necessary for reactions in acidic or basic solution, respectively.

8. Balance the oxygen atoms by adding H_2O molecules to one side of the equation.

9. Check to make sure that the equation is balanced in both atoms and total charges.

Single displacement reactions occur when a metal reacts with an acid or another metal salt; such a reaction either dissolves the metal and forms a precipitate or generates hydrogen gas. This reaction can be predicted from the activity series, with metals able to reduce any metallic compound below it in the series. The series is divided into active metals, which are those near the top, and inert metals, which are those near the bottom.

4.9. Quantitative Analysis Using Titrations

Quantitative analysis is a method by which the amounts or concentrations of substances in a sample are determined. One of the most common methods in quantitative analysis is the titration, in which a measured volume of a solution, the titrant, is added to another solution in order to determine its concentration. Before analyzing an unknown, most titrants are standardized, which means they are used to titrate a solution of an accurately known concentration. From this reaction, the concentration of the titrant is calculated, and then the unknown is analyzed. The equivalence point is the point where just enough titrant has been added to completely react with the sample. It's common to follow the progress of a titration by using a visual indicator, particularly in the case of an acid–base titration. The color change in the indicator occurs at the endpoint, which is generally close to the equivalence point.

Self-Test

1. Cations and anions that are surrounded by a shell of water molecules are called
   _____
   A. solutes.
   B. hydrated ions.
   C. solvents.
   D. electrolytes.

2. What is the molarity of a solution if 5 g of NaOH is dissolved in 500 mL of water?
   A. 0.12 M
   B. 0.01 M
   C. 0.25 M
   D. 0.07 M

3. How many grams of $CuSO_4$ are required to make 750 mL of a 2.5 M solution?
   A. 299
   B. 399
   C. 250
   D. 350

4. What volume of 12 M HCl is required to make 4 L of 0.5 M HCl?
   A. 333 mL
   B. 144 mL
   C. 167 mL
   D. 250 mL

5. Barium chloride and sodium sulfate react to produce sodium chloride and barium sulfate. If 100 mL of a 2 M solution of sodium sulfate is used, how many grams of barium chloride are needed for the reaction to go to completion?
   A. 17.23
   B. 20.8
   C. 34.47
   D. 41.6

6. Copper II nitrate reacts with ammonium sulfide to produce a copper II sulfide precipitate. How many grams of copper II sulfide can be formed from mixing 10 mL of 1 M copper II nitrate with excess ammonium sulfide?
   A. 0.5
   B. 1
   C. 1.5
   D. 2

7. Balance the following net ionic equation:

$$CO_3{}^{2-} + 2\,H_3O^+ \rightarrow \underline{\quad} H_2O + \underline{\quad} H_2CO_3$$

   A. 1,2
   B. 2,1
   C. 2,2
   D. 1,1

8. Balance the following complete ionic equation:

$$Cu + 4HNO_3 \rightarrow \underline{\quad} Cu^{2+} + \underline{\quad} H_2O + \underline{\quad} NO_3{}^- + \underline{\quad} NO_2$$

   A. 1,1,2,2
   B. 1,2,1,2
   C. 1,2,2,2
   D. 1,2,2,1

9. A precipitation reaction belongs to what general class of reactions?
   A. Oxidation-reduction reaction
   B. Acid–base reaction
   C. Condensation reaction
   D. Exchange reaction

10. Vinegar has an approximate pH of 5.0, which makes it _____
   A. an acid, because of its low pH.
   B. a base, because of its low pH.
   C. an acid because of its high pH.
   D. a base because of its high pH.

11. Ammonia has an approximate pH of 11.0, which makes it _____
   A. an acid, because of its low pH.
   B. a base, because of its low pH.
   C. an acid because of its high pH.
   D. a base because of its high pH.

12. When an acid and a base are mixed together, one of the products is always water. What kind of reaction is this?
   A. Precipitation reaction
   B. Complexation reaction
   C. Neutralization reaction
   D. Color formation reaction

13. What is the pH of a solution with a hydrogen ion concentration of 3.98×10^{-4}?
    A. -3.4
    B. 3.4
    C. 4
    D. 10.6

14. What is the hydrogen ion concentration of a solution with a pH of 8.6?
    A. 3×10^{-9}
    B. 3×10^{-8}
    C. 3×10^{-10}
    D. 3×10^{-12}

15. Which of the following is not a source of sulfur dioxide in the atmosphere?
    A. Combustion of fossil fuels
    B. Volcanoes
    C. Forest fires
    D. Anaerobic respiration

16. The primary acids responsible for acid rain are _____
    A. hydrochloric acid and nitric acid.
    B. sulfuric acid and phosphoric acid.
    C. nitric acid and sulfuric acid.
    D. sulfuric acid and carbonic acid.

17. Complete this ion transfer equation:
$$Cu^{2+} + Fe \rightarrow \_\_ + \_\_$$
    A. $CuFe + 2^{e-}$
    B. $Cu + 2Fe$
    C. $Cu^{2-} + Fe^{2+}$
    D. $Cu + Fe^{2+}$

18. Which of the following has the highest activity?
    A. Al
    B. Cr
    C. Li
    D. Hg

19. In a titration, the solution of a known concentration that is added to a sample of an unknown concentration is called the _____
    A. equivalence point.
    B. endpoint.
    C. titrant.
    D. standard solution.

20. What volume of 0.1 M NaOH must be added to reach the equivalence point in a solution of 50 mL of 0.15 M HCl?

 A. 33 mL

 B. 75 mL

 C. 50 mL

 D. 61 mL

Answers: 1. B; 2. C; 3. A; 4. C; 5. B; 6. B; 7. B; 8. C; 9. D; 10. A; 11. D; 12. C; 13. B; 14. A; 15. D; 16. C; 17. D; 18. C; 19. C; 20. B

CHAPTER 5

Energy Changes in Chemical Reactions

Key Words

bomb calorimeter	enthalpy of reaction	radiant energy
calorie	enthalpy of solution	specific heat
calorimetry	enthalpy of vaporization	standard conditions
carbon cycle	exothermic	standard enthalpy of
change in enthalpy	greenhouse effect	formation
chemical energy	greenhouse gases	standard enthalpy of
closed system	heat	reaction
coal	heat capacity	standard state
constant-pressure	Hess's law	state function
calorimeter	isolated system	surroundings
electrical energy	joule	syngas
endothermic	kinetic energy	system
energy	mechanical work	temperature
enthalpy	molar heat capacity	thermal energy
enthalpy of combustion	nuclear energy	thermochemistry
enthalpy of formation	open system	
enthalpy of fusion	potential energy	

By the end of this chapter, you should be able to:

- Understand the different forms of energy
- Understand how energy, work, and heat are related
- Describe a state function
- Calculate enthalpy changes in reactions using Hess's law and thermochemical cycles
- Understand calorimetry experiments and do the calculations derived from them
- Understand how the calorie content in your food is determined
- Describe the use of fossil fuels for energy
- Understand the carbon cycle
- Describe the greenhouse effect

Chapter Overview

In the previous discussions of chemical reactions, we've only been concerned about the atoms and compounds involved and the products that are formed from them. While this is a good way to start, it is not the complete way to look at a chemical reaction. All chemical reactions either give off energy in the form of heat and/or light, or they take in energy from their surroundings. This chapter looks at thermochemistry, which describes the energy changes that occur during chemical reactions.

5.1. Energy and Work

Energy is defined as the ability to do work, and it exists in several forms. The law of conservation of energy states that the amount of energy in the universe remains constant, which means that energy has the ability to change from one form to another. The forms of energy currently known to exist are radiant energy, from the light spectrum; thermal energy, from the vibration of atoms and molecules; chemical energy, from the making and breaking of bonds during a chemical reaction; nuclear energy, from nuclear reactions; and electrical energy, from the flow of charged particles. Energy that is stored in an object and related to the position of that object is known as *potential energy.* Energy that is based on the motion of an object is *kinetic energy.* Energy is most often defined by its effect on matter, either in terms of mechanical work or heat. Heat is the transfer of thermal energy from a hotter object to a cooler one. The most common units of energy are the joule, defined as $1 \ (kg \times m^2)/s^2$, and the calorie, defined as the amount of energy needed to raise the temperature of 1 gram of water by 1°C.

5.2. Enthalpy

In order to study the energy flow that occurs during a chemical reaction, we have to distinguish between what we are studying and everything around what we are studying. The system is the small part of the universe we are interested in, such as a chemical reaction. The surroundings are everything else in the universe, including the vessel in which our reaction is taking place. An open system is one that can exchange both matter and energy with its surroundings. A closed system can exchange only energy with the surroundings. And an isolated system can exchange neither matter nor energy with its surroundings. Any process that gives off heat from the system to the surroundings is said to be *exothermic,* while any process that takes heat in from the surroundings into the system is said to be *endothermic.* *Enthalpy* is a measure of the heat transferred from a system or taken in by a system from the surroundings at constant pressure; it is a state function, meaning that it represents only the final state of a system, regardless of the path taken to reach that state. Enthalpy is measured as the change in enthalpy, ΔH. If ΔH is negative, the reaction is exothermic and heat is transferred from the system into the surroundings. If ΔH is positive, the reaction is endothermic and heat is taken into the system from the surroundings.

For a chemical reaction, the enthalpy of reaction (ΔH_{rxn}) is the enthalpy difference between the products and the reactants. The magnitude of ΔH_{rxn} depends on the physical state of the reactants and products, with changes in physical state being accompanied by their own enthalpies:

Enthalpy of combustion, ΔH_{comb}: Enthalpy changes have been measured for the combustion of virtually any substance that will burn in O_2; these values are usually reported as the enthalpy of combustion per mole of substance.

Enthalpy of fusion, ΔH_{fus}: The enthalpy change that accompanies the melting, or fusion, of 1 mol of a substance; these values have been measured for almost all of the elements and for most simple compounds.

Enthalpy of vaporization, ΔH_{vap}: The enthalpy change that accompanies the vaporization of 1 mol of a substance; these values have also been measured for nearly all the elements and for most volatile compounds.

Enthalpy of solution, ΔH_{soln}: The enthalpy change when a specified amount of solute dissolves in a given quantity of solvent.

Enthalpy of formation, ΔH_f: The enthalpy change for the formation of 1 mol of a compound from its component elements.

These enthalpies can be used along with Hess's law, which states that the overall change in enthalpy for a reaction is the sum of the changes in enthalpy for the individual reactions. This law can be used to calculate the ΔH for a reaction by following these steps:

1. Identify the equation whose ΔH value is unknown, and write individual reactions with known ΔH values that, when summed, will give the desired equation.

2. Arrange the chemical equations so that the reaction of interest is the sum of the individual reactions.

3. If a reaction must be reversed, change the sign of ΔH for that reaction. Additionally, if a reaction must be multiplied by a factor to obtain the correct number of moles of a substance, multiply its ΔH value by that same factor.

4. Sum the individual reactions and their corresponding ΔH values to obtain the reaction of interest and the unknown ΔH.

Most tabulated enthalpy data will be given under a set of standard conditions: a pressure of 1 atm, a concentration of 1 M for all solutions, and each species present in its most stable form at a pressure of 1 atm and a temperature of 25°C. The standard enthalpy of formation, $\Delta H°_f$, is given under these conditions, and the standard enthalpy of reaction, $\Delta H°_{rxn}$, can be calculated from these standard enthalpies of formation.

5.3. Calorimetry

Enthalpy changes can be measured in the lab and can be calculated. *Calorimetry* is that set of techniques used in the lab to make these measurements. It uses a device called a *calorimeter* to measure the change in temperature of either a system or the surroundings when a chemical reaction is carried out. This change will be dependent upon the amount of heat transferred by the reaction and upon the heat capacity of the system. The heat capacity of something is the amount of energy needed to raise its temperature by 1°C. Two related measures are specific heat, which is the energy needed to raise 1 gram of a substance by 1°C, and the molar heat capacity, which is the amount of energy needed to raise 1 mol of a substance by 1°C. Calorimetry experiments may be carried out in either a constant-pressure calorimeter or a bomb calorimeter. A bomb calorimeter operates at constant volume and is especially useful for measuring combustion reactions, such as those reactions used to establish the calorie content in food.

5.4. Thermochemistry and Nutrition

The calorie that is reported on food packages is the kcal, and it is determined through the combustion of a sample of the food in a bomb calorimeter. This is used even though the processes in our bodies that break down food don't always generate the same products as a straight combustion. This holds especially true for proteins and other nitrogen-containing foods, which actually break down in the body to produce urea. The usual calorie breakdowns for foods are 9 Cal per gram for fats and 4 Cal per gram for proteins and carbohydrates.

5.5. Energy Sources and the Environment

Energy usage in the United States accounts for about 30% of the world's energy consumption, though our population accounts for only about 5% of the world's population. About 80% of the worldwide energy comes from the combustion of fossil fuels. Of these fossil fuels, natural gas and petroleum are the preferred sources due to their availability and ease of transport and refinement. Coal is perhaps the most available fossil fuel but is far more difficult to transport and burn, making it a much less attractive fuel source, as does the fact that it yields the least amount of energy per gram of the fossil fuels. Combustion of any organic compound yields carbon dioxide, and burning fossil fuels have released a large amount of CO_2 into the atmosphere. It is believed that this has upset the earth's carbon cycle, leading to an increasing amount of CO_2 in the atmosphere. Carbon dioxide is a greenhouse gas that traps heat before it can be released into space; it is believed that the increasing levels of CO_2 in the atmosphere have caused an effect known as the *greenhouse effect,* resulting in a slow rise in the earth's temperature. There is controversy surrounding the role that fossil fuels play in this effect and argument that the increase seen in recent history is negligible compared to other temperature fluctuations that have occurred throughout the earth's history.

Self-Test

1. Which of the following is not a form of energy?
   A. Radiant energy
   B. Electrical energy
   C. Magnetic energy
   D. Thermal energy

2. The statement that "the total amount of energy in the universe remains constant" is known as _____
   A. the law of kinetic energy.
   B. the law of conservation of energy.
   C. Hess's law.
   D. the law of mechanical work.

3. What is the kinetic energy of a 1360 kg car traveling at a speed of 31.3 m/s?
    A. 13.4×10^5 (kg m²)/s²
    B. 8.9×10^5 (kg m²)/s²
    C. 6.7×10^5 (kg m²)/s²
    D. 11.5×10^5 (kg m²)/s²

4. What is the potential energy of that same car if it is perched on a ledge 50 meters above ground level?
    A. 13.4×10^5 J
    B. 8.9×10^5 J
    C. 6.7×10^5 J
    D. 11.5×10^5 J

5. A system that can exchange neither energy nor matter with the surroundings is known as a(n) _____
    A. open system.
    B. closed system.
    C. isolated system.
    D. state function.

6. An ice cube melts to form liquid water. The enthalpy associated with this change is known as the _____
    A. enthalpy of reaction.
    B. enthalpy of solution.
    C. enthalpy of formation.
    D. enthalpy of fusion.

7. What is the standard enthalpy of formation, $\Delta H°_f$, for hydrogen gas?
    A. 1273 kJ
    B. −1717 kJ
    C. 5 kJ
    D. 0 kJ

8. The enthalpy of solution for ammonium nitrate is 25.7 kJ/mole, making the reaction _____
    A. endothermic.
    B. exothermic.

9. A sample of 15 mL bleach ($NaClO$) and 15 mL sodium sulfite (Na_2SO_3) is placed in a calorimeter and measures a temperature change of 7°C. This reaction is _____
    A. endothermic.
    B. exothermic.

10. A calorimetry experiment has a total volume (water and reaction) of 80 mL and produces a 3°C temperature change (ΔT). Assume that the density and the specific heat (C_p) of the experiment are both equal to 1. Calculate the heat of reaction ($\Delta H = vdC_p\Delta T$).

   A. 2.40 cal

   B. 240 cal

   C. −2.40 cal

   D. −240 cal

11. A calorimetry experiment has a total volume (water and reaction) of 69 mL and produces a −2°C temperature change (ΔT). Assume that the density and the specific heat (C_p) of the experiment are both equal to 1. Calculate the heat of reaction.

   A. 138 cal

   B. 1.38 cal

   C. −138 cal

   D. −1.38 cal

12. An enthalpy of combustion is generally best measured in a(n) _____

   A. constant-pressure calorimeter.

   B. coffee cup calorimeter.

   C. closed calorimeter.

   D. bomb calorimeter.

13. The nutritional calorie is equal to _____

   A. 10 J.

   B. 100 J.

   C. 1000 J.

   D. 10,000 J.

14. The caloric content in foods is generally measured in a(n) _____

   A. constant-pressure calorimeter.

   B. coffee cup calorimeter.

   C. closed calorimeter.

   D. bomb calorimeter.

15. Which type of food has the highest caloric content?

   A. Proteins

   B. Carbohydrates

   C. Fats

   D. Vegetables

16. The combustion of glucose produces 15.6 kJ/g of energy. How many grams of glucose would a 70 kg person need to burn to climb a 150 m high hill?

    A. 6.6 g

    B. 3.7 g

    C. 13.2 g

    D. 24.6 g

17. Which of the following is not a type of coal?

    A. Anthracite

    B. Bituminous

    C. Charcoal

    D. Lignite

18. The top source of energy in the United States comes from _____

    A. coal.

    B. natural gas.

    C. nuclear power.

    D. oil.

19. The gas primarily responsible for the greenhouse effect is _____

    A. CO_2.

    B. CO.

    C. NO_2.

    D. NO.

20. The distribution and flow of carbon throughout the planet is known as the _____

    A. greenhouse effect.

    B. carbon cycle.

    C. carbon dioxide cycle.

    D. thermal cycle.

Answers: 1. C; 2. B; 3. C; 4. C; 5. C; 6. D; 7. D; 8. A; 9. B; 10. B; 11. C; 12. D; 13. C; 14. D; 15. C; 16. A; 17. C; 18. D; 19. A; 20. B

CHAPTER 6

The Structure of Atoms

Key Words

absorption spectrum
amplitude
atomic orbital
Aufbau principle
azimuthal quantum
  number
blackbody radiation
d block
degenerate
effective nuclear charge
electromagnetic radiation
electron shielding
electron spin
emission spectrum
excited state
f block

frequency
ground state
Heisenberg uncertainty
  principle
Hund's rule
line spectrum
magnetic quantum
  number
nodes
Pauli exclusion principle
p block
periodic
photoelectric effect
photoelectron
photons
principal quantum number

principal shell
quantum
quantum mechanics
quantum numbers
s-block elements
spectroscopy
speed
speed of light
standing wave
subshell
valence electrons
wave
wave function
wavelength
wave-particle duality

By the end of this chapter, you should be able to:

- Explain the characteristics of electromagnetic energy
- Understand the electromagnetic spectrum
- Understand how energy is quantized
- Understand how the electronic structure of atoms gives rise to their atomic spectra
- Understand the concept of the wave-particle duality of matter
- Understand how quantum mechanics applies to chemistry
- Write the electron configuration of an element
- Relate an element's electron configuration to its position in the periodic table

41

Chapter Overview

In previous chapters we've talked about the fundamental concepts of atoms and molecules. Now we build on that foundation and come to a deeper understanding of the electronic structure of the atom. From this, you will better understand reaction stoichiometries and the shapes of chemical compounds. You will also begin to see some of the patterns and usefulness of the periodic table.

6.1. Waves and Electromagnetic Radiation

It is necessary to have a foundational knowledge of electromagnetic radiation and waves in order to understand the electronic structure of the atom. Energy moves through space in a periodic oscillation known as a *wave*. Each wave has a characteristic wavelength, which is the distance between waves. Waves are also known by a frequency, which is the number of waves that pass through a fixed point during a given span of time. A wave's speed describes how fast it moves through space, while the amplitude describes the magnitude of the oscillation. All waves are periodic, meaning that they repeat regularly in both time and space. When an electric wave and a magnetic wave combine in a perpendicular fashion, it gives rise to electromagnetic radiation, which is known to move at the speed of light. Electromagnetic radiation falls into a spectrum of frequencies and wavelengths and includes such radiation as visible light, X-rays, gamma rays, microwaves, and radio waves.

6.2. The Quantization of Energy

Modern physics principles arise primarily from observations that could not be explained by classical physics, such as the radiation given off by hot objects (blackbody radiation). Max Planck came up with the idea that energy can only be absorbed and emitted in integral multiples, which he called a *quantum*. The energy of a quantum is proportional to the frequency of the radiation through a constant now known as *Planck's constant* (h). Planck's ideas were expanded by Albert Einstein to explain why certain metals eject electrons when exposed to light. Einstein named this the *photoelectric effect*. Einstein also theorized the existence of photons, or particles of light with a particular energy.

6.3. Atomic Spectra and Models of the Atom

The atoms of an element emit light at very specific wavelengths, giving what is known as a *line spectrum*. Niels Bohr generated the line spectrum of the hydrogen atom and theorized that the electrons in the atom moved in circular orbits with certain radii. As the electrons transitioned from higher, excited orbits to the lowest, ground-state orbit, the spectrum was emitted. Bohr's model, however, could not explain the spectra of elements heavier than hydrogen.

The majority of light is composed of many wavelengths. When light is only composed of a single wavelength, it is called *monochromatic*. Atoms are also capable of absorbing light, resulting in a transition from the ground state, or a lower excited state, to a higher excited state. This gives rise to what is known as an *absorption spectrum*. This spectrum is the opposite of the emission spectrum and has dark lines in the same positions as the light lines in an emission spectrum.

6.4. The Relationship between Energy and Mass

A fundamental concept in modern physics is that the electron possesses the properties of both a wave and a particle, which is known as *wave-particle duality*. Louis de Broglie found that the wavelength of a particle is equal to Planck's constant divided by the product of the mass and velocity of the particle. The orbits of electrons can be described by the harmonics of their vibrations. The lowest-energy wave is the fundamental vibration, with higher-energy vibrations being overtones with an increasing number of nodes at which the wave passes through zero. Heisenberg's uncertainty principle says that it is impossible to precisely know both the speed and the location of an electron.

6.5. Atomic Orbitals and Their Energies

Quantum mechanics uses wave functions to describe the mathematical relationship between the motion of electrons in atoms and molecules and their energies. Wave functions possess five properties:

1. A wave function uses three variables to describe the position of an electron.

2. The magnitude of the wave function at a particular point in space is proportional to the amplitude of the wave at that point.

3. The square of the wave function at a given point is proportional to the probability of finding an electron at that point, which leads to a distribution of probabilities in space.

4. Describing the electron distribution as a standing wave leads to a set of "quantum numbers" characteristic of each wave function.

5. Each wave function is associated with a particular energy.

There are four quantum numbers that describe the energy and spatial distribution of an electron. The first quantum number is the principal quantum number n. This number is an integer that describes in which shell the electron is located. The second quantum number, l, is the azimuthal quantum number. This number is an integer between 0 and $n - 1$, and it describes the shape of the electron distribution. The third quantum number is the magnetic quantum number m_l. This number can have $2l + 1$ integer values, ranging from $-l$ to $+l$, and it describes the orientation of the electron distribution. The final quantum number will be defined in the next section.

6.6. Building up the Periodic Table

The fourth quantum number is the electron spin quantum number, m_s. This number can have a value of either $+1/2$ or $-1/2$, and it describes the two possible orientations of a magnet in a magnetic field. Several principles and rules can be understood through the quantum numbers and can be used to describe patterns in the periodic table. The Pauli exclusion principle states that no orbital can have more than two electrons. The Aufbau principle says that orbitals fill from the lowest energy and work their way up. Finally, Hund's rule states that the lowest-energy arrangement of electrons is the one that places them in degenerate orbitals with their spins parallel. Taken together, these rules allow us to construct unique electron configurations for each element in the periodic table and to establish patterns in the periodic table. In terms of chemical reactions, the most important electrons are those located in the outermost, or valence, shell of an atom. The arrangement of elements in the periodic table results in blocks corresponding to the filling of the *ns*, *np*, *nd*, and *nf* orbitals to produce the unique chemical properties of the *s*-block, *p*-block, *d*-block, and *f*-block elements.

Self-Test

1. Which of the following is not electromagnetic radiation?
   A. Microwaves
   B. Gamma rays
   C. Alpha particles
   D. Radio waves

2. What is the wavelength of a 4.8×10^8 wave?
   A. 6.25 m
   B. 0.625 m
   C. 0.0625 m
   D. 0.00625 m

3. Which of the following wavelengths falls in the red range of the visible spectrum?
   A. 425 nm
   B. 550 nm
   C. 600 nm
   D. 675 nm

4. Planck's constant is equal to _____
   A. 6.626×10^{-34} J · s.
   B. 3×10^8 m/s.
   C. 6.02×10^{23} photons/mole.
   D. 3.414×10^{-15} J · s.

5. The ejection of an electron in a metal atom that is exposed to light is known as the _____
   A. excited state.
   B. photoelectric effect.
   C. blackbody radiation.
   D. line spectrum.

6. What is the energy of a photon with a frequency of 10^6 Hz?
   A. 6.626×10^{-40} J
   B. 6.626×10^{-28} J
   C. 6.626×10^{-15} J
   D. 6.626×10^{-12} J

7. When an excited electron decays to a lower energy state, it produces a(n) _____
   A. line spectrum.
   B. continuous spectrum.
   C. emission spectrum.
   D. absorption spectrum.

8. When the emitted light of an atom is passed through a prism, it produces a(n) _____
   A. line spectrum.
   B. continuous spectrum.
   C. emission spectrum.
   D. absorption spectrum.

9. Which element would be responsible for the blue color in a firework?
   A. Strontium
   B. Sodium
   C. Barium
   D. Copper

10. Which of the following is not used in neon signs?
    A. Neon
    B. Barium
    C. Mercury
    D. Sodium

11. How many nodes would a third overtone have?
    A. One
    B. Two
    C. Three
    D. Four

12. What is the wavelength of the second overtone of a string that is 24 cm long?
    A. 12 cm
    B. 8 cm
    C. 24 cm
    D. 16 cm

13. What is the minimum uncertainty in the position of a 60 g ball moving at a speed of 0.75 m/s?

    A. 1.25×10^{-34} m

    B. 1.11×10^{-34} m

    C. 1.16×10^{-34} m

    D. 3.3×10^{-34} m

14. The p orbital has what kind of shape?

    A. Spherical

    B. Dumbbell

    C. Clover

    D. Polygonal

15. How many subshells does a d orbital have?

    A. One

    B. Two

    C. Three

    D. Four

16. Which quantum number describes the orientation of the electron distribution?

    A. Principal quantum number

    B. Azimuthal quantum number

    C. Magnetic quantum number

    D. Electron spin quantum number

17. The statement that no two electrons in an atom can have identical sets of quantum numbers is known as _____

    A. the Pauli exclusion principle.

    B. the Aufbau principle.

    C. Hund's rule.

    D. the octet rule.

18. What is the electron configuration for Pb?

    A. $[Xe]6s^2 4f^{14} 5s^2 5d^{10}$

    B. $[Xe]4f^{14} 5s^2 5p^2 5d^{10}$

    C. $[Xe]4f^{14} 5s^2 5p^6 5d^6$

    D. $[Xe]6s^2 4f^{14} 5d^{10} 6p^2$

19. Which element has the configuration $[Kr]5s^2 4d^{10} 5p^5$?

    A. Sb

    B. Te

    C. I

    D. Xe

20. What is the set of quantum numbers for the valence electron in Na?
    A. $3, 0, 0, -\frac{1}{2}$
    B. $3, 0, 0, +\frac{1}{2}$
    C. $3, 1, 0, -\frac{1}{2}$
    D. $3, 1, 0, +\frac{1}{2}$

Answers: 1. C; 2. B; 3. D; 4. A; 5. B; 6. B; 7. C; 8. A; 9. D; 10. B; 11. B; 12. D; 13. C; 14. B; 15. C; 16. C; 17. A; 18. D; 19. C; 20. B

CHAPTER 7

The Periodic Table and Periodic Trends

Key Words

actinides	fullerenes	molar volume
alkali metals	group transfer reaction	nanotubes
alkaline earths	halogens	noble gases
amplification mechanism	hyponatremia	octaves
chalcogens	ionic radius	pnicogens
congeners	ionization energy	pseudo-noble gas
covalent atomic radius	ion pump	configuration
electron affinity	isoelectric series	transition metals
electronegativity	lanthanides	triads
essential trace elements	main group elements	van der Waals radius

---

By the end of this chapter, you should be able to:

- Have an overview of the history of the periodic table
- Understand the periodic trends of atomic radii
- Predict relative ionic sizes within an isoelectronic series
- Correlate ionization energies and electron affinities with the chemistry of the elements
- Understand how the electronegativity of an element influences its chemistry
- Understand the relationship between an element's position in the periodic table and its chemical properties
- Describe some of the workings of trace elements in biological systems and the symptoms of some of their deficiencies

Chapter Overview

In the last chapter, we talked quite a bit about quantum mechanics and the electron configurations of the elements. These configurations follow definite patterns that influence the properties of the elements and the reactions that they can undergo. The elements can be ordered by these patterns, and this chapter focuses on discussing the periodic table of the elements and the patterns that are contained within it.

7.1. The History of the Periodic Table

Today's periodic table is the result of a lengthy process of attempts by scientists to place the elements into patterns. One of the first attempts came in the early 1800s when Johannes Dobereiner noticed that many elements could be grouped into triads with similar properties. John Newlands postulated that the elements could be grouped into octaves, much like a musical scale. Our modern periodic table came from the work of Julius Lothar Meyer and Dimitri Mendeleev, who focused on atomic mass to establish the pattern of the elements. The table was further refined by H. G. J. Moseley, who established that the pattern was based upon atomic numbers instead of masses. The current version of the table arranges the elements according to their electron configuration, with elements in the same columns having the same valence electron configuration. This arrangement allows us to see patterns that affect the types of chemical reactions that the elements undergo and the chemical compounds that can be formed.

7.2. Sizes of Atoms and Ions

There are a number of ways to define the size of an atom or ion, depending upon what it's bonded to. The covalent atomic radius is half the internuclear distance of a molecule that contains two identical ions that are bonded to each other. The metallic atomic radius is half the internuclear distance of two adjacent atoms in a metallic element. The van der Waals radius is half the internuclear distance of two nonbonded atoms in a solid. The atomic radii increase from right to left across a row and from top to bottom down a column. In an ionic compound, the repulsions between electrons result in ionic radii that can be either larger or smaller than the parent atom. The sizes of atoms or molecules that have the same number of electrons but different nuclear charges can be compared through an *isoelectronic series;* such a series shows a correlation between increasing nuclear charge and decreasing size.

7.3. Energetics of Ion Formation

The most important thing in determining the kind of compounds that an element can form is that element's tendency to lose or gain electrons. The energy required to remove an electron from an atom in the gaseous state is known as the *ionization energy.* The ionization energy increases as you try to remove successive electrons, and the most energy is needed to remove electrons from a filled inner shell. In chemical reactions, only the valence electrons are removed. Ionization energy increases diagonally from the periodic table's lower left to the upper right. This trend has minor deviations that occur when either the parent atom or the ion attain a particularly stable electron configuration known as a *pseudo-noble gas configuration.* The opposite of ionization energy is electron affinity, which is the energy change that occurs when an electron is added to a gaseous atom. Elements with the highest affinity for added atoms are those with the highest ionization energy and the smallest size; these can be found in the periodic table's upper right-hand corner. An element's electronegativity is its ability to attract electrons to itself in a compound. The most electronegative elements appear in the periodic table's upper right-hand corner. Elements with high electronegativities tend to be nonmetals and electrical insulators and tend to act as oxidants in reactions. Elements with lower electronegativities tend to be metals and electrical conductors and tend to act as reductants in reactions.

7.4. The Chemical Families

Elements that have the same valence electron configuration have similar chemistry and are grouped into chemical families. Group 1 is known as the *alkali metals* and has a valence configuration of ns^1. Group 2 is known as the *alkaline earths* and has an ns^2 valence configuration. The transition metals make up Groups 3–10, and these elements contain partially filled d orbitals. Group 13 elements have configurations of ns^2np^1, and Group 14 elements have configurations of ns^2np^2. Group 15 elements are known as *pnicogens* and have an ns^2np^3 valence configuration. Group 16 elements are known as *chalcogens* and have an ns^2np^4 configuration. Group 17 elements are the halogens and have a configuration of ns^2np^5. The noble gases are Group 18 and have a filled shell configuration of ns^2np^6. The lanthanides and the actinides have filling f orbitals. The number of electrons in the valence shell determines the oxidation states and chemical reactions of an element. Noble gases tend to be quite unreactive due to the filled valence shells. Elements with mostly filled valence shells are most likely to accept electrons, while those with only partially filled valence shells are most likely to lose their electrons. Elements that have approximately half-filled valences are likely to either lose their electrons or gain additional ones.

7.5. Trace Elements in Biological Systems

Approximately 28 of the currently known elements are essential for the biological function of a species, and 19 are essential to humans. Many of these are known as *essential trace elements* because they are only needed in very small amounts. Even though they are trace elements, they are necessary for important biological functions and accomplish this by participating in amplification mechanisms. Macrominerals are essential in larger amounts and are needed for structural elements and electrolytes. These minerals are transferred into and out of cells by ion pumps, and imbalances in their concentrations can be dangerous or lethal. Other trace elements are needed to catalyze biological oxidation-reduction reactions or are components of biological molecules.

Self-Test

1. The modern periodic table is largely attributed to the work of _____
   A. Johannes Dobereiner.
   B. John Newlands.
   C. Dimitri Mendeleev.
   D. H. G. J. Moseley.

2. Who was the scientist who attributed the pattern of the elements to the atomic number?

    A. Johannes Dobereiner

    B. John Newlands

    C. Dimitri Mendeleev

    D. H. G. J. Moseley

3. Calculate the molar volume of gallium, given that it has a density of 5.91 g/cm^3.

    A. 9.6 cm^3/mol

    B. 11.8 cm^3/mol

    C. 13.2 cm^3/mol

    D. 15.3 cm^3/mol

4. Which element would have the largest atomic radius?

    A. Li

    B. B

    C. N

    D. F

5. Which element would have the smallest atomic radius?

    A. Ba

    B. Sr

    C. Ca

    D. Mg

6. The radius that describes the internuclear distance between two nonbonded atoms in a solid is known as the _____

    A. covalent atomic radius.

    B. ionic radius.

    C. metallic atomic radius.

    D. van der Waals radius.

7. Which of the following elements would have the highest ionization energy?

    A. Pb

    B. I

    C. S

    D. Ne

8. Which of the following elements would have the lowest ionization energy?

    A. Ni

    B. Ba

    C. Se

    D. Zr

9. Which of the following elements would have the most negative electron affinity?

    A. Hg

    B. Ir

    C. At

    D. Pb

10. Which of the following elements would have the least negative electron affinity?

    A. Co

    B. Zn

    C. Ge

    D. Br

11. Which of the following elements would have the highest electronegativity?

    A. Ag

    B. W

    C. Cl

    D. As

12. Which of the following elements would have the lowest electronegativity?

    A. Cd

    B. I

    C. Fr

    D. Os

13. Which of the following elements is a pnicogen?

    A. Br

    B. P

    C. Al

    D. Ca

14. Which of the following elements is a chalcogen?

    A. Te

    B. Ga

    C. Rh

    D. Rb

15. Which of the following is not an allotrope of carbon?

    A. Diamond

    B. Graphite

    C. Fullerenes

    D. Coal

16. The alkali earths form which ion?

    A. M^{1+}

    B. M^{2+}

    C. M^{1-}

    D. M^{2-}

17. Which of the following would not be an oxidation state of a pnicogen?

    A. -4

    B. -1

    C. $+2$

    D. $+5$

18. A lack of iodine can lead to a condition known as _____

    A. scurvy.

    B. hyponatremia.

    C. goiter.

    D. hyperthyroidism.

19. Ions are selectively transported through cell membranes by the _____

    A. amplification mechanism.

    B. ion pump.

    C. macrominerals.

    D. group-transfer reaction.

20. Which of the following is not an essential element for humans?

    A. Oxygen

    B. Magnesium

    C. Zinc

    D. Lithium

Answers: 1. C; 2. D; 3. B; 4. A; 5. D; 6. D; 7. D; 8. B; 9. C; 10. B; 11. C; 12. C; 13. B; 14. A; 15. D; 16. B; 17. A; 18. C; 19. B; 20. D

CHAPTER 8

Structure and Bonding I: Ionic versus Covalent Bonding

Key Words

bond distance	electron-deficient	Lewis acid
bond energy	molecules	Lewis base
bonding pair	enthalpy of sublimation	Lewis dot structure
bond order	expanded-valence	lone pair
Born–Haber cycle	molecules	melting point
chemical bond	formal charge	nonpolar
coordinate covalent	hardness	octet rule
bond	ionic bond	polar covalent bond
covalent bond	lattice	resonance structures
dipole moment	lattice energy	sublimation

By the end of this chapter, you should be able to:

- Understand what is involved in an ionic bond
- Understand how lattice energy influences the properties of an ionic compound
- Understand and use a Born–Haber cycle
- Draw Lewis dot structures for elements and compounds
- Understand resonance structures
- Define Lewis acids and bases
- See the pattern between bond order, bond length, and bond energy
- Calculate the percent ionic character of a covalent polar bond

Chapter Overview

The previous two chapters talked about atoms, elements, and their properties. This chapter takes the next logical step and begins the study of bonding those atoms and elements together to form chemical compounds. The overall goal is to provide you with an understanding of how the properties of the parts (the atoms) influence the structure and properties of the whole (the compound).

8.1. An Overview of Chemical Bonding

In Chapter 2 we spoke of a chemical bond as the force that holds together the atoms in a chemical compound. These bonds are classified as *covalent* when the electrons are shared between the participating atoms or as *ionic* when positive and negative ions are held together by electrostatic forces. All bonds share certain characteristics: (1) the compounds are more stable and have lower energy than the atoms that compose them; (2) a certain amount of energy—called the *lattice energy* or *bond energy*—is required to break the bonds of a compound; and (3) each bond has a characteristic bond distance.

8.2. Ionic Bonding

Two things influence the electrostatic forces that make an ionic bond: the charges on the ions and the distance between them. As the charges increase, so does the attraction. As the distance (internuclear) between them decreases, the attraction increases. The total energy of an ionic compound is a balance between the attractive and repulsive forces of the ions. As ions are brought together to form a compound, energy is released. The energy then needed to break this bond is referred to as the *bond energy*.

8.3. Lattice Energies in Ionic Solids

Previously you learned that compounds arranged in a lattice structure have certain distinct properties, such as a high stability. The lattice energy, *U*, represents the amount of energy it would take to break an ionic solid into its gaseous ions. This energy is proportional to the charges on the ions, is inversely proportional to the distance between the ions, and is always a positive number. Since the energy is based upon the product of the charges, increasing the charge of an ion will double or triple the lattice energy. Lattice energies are therefore highest for substances that contain small, highly charged ions. Increasing lattice energies lead to an increase in hardness and melting point and a decrease in solubility. The lattice energy is the most important factor in determining the stability of an ionic compound.

Lattice energies cannot be measured directly but must be measured through the Born–Haber cycle, in which the compound of interest is formed in steps from its component elements. The Born–Haber cycle makes use of Hess's law to calculate the lattice energy from the measured enthalpy of formation of the compound.

8.4. Introduction to Lewis Electron Structures

Lewis dot symbols are a convenient way to predict the number of bonds that can be formed by most elements in their compounds. To write a Lewis dot symbol, we place up to eight dots around the chemical symbol of an element, with a pair each above, below, and to the left and right of the symbol. The octet rule states that atoms tend to lose, gain, or share electrons to reach a total of eight valence electrons. The exceptions to this rule are hydrogen and helium, which have full valence shells with only two electrons.

8.5. Lewis Structures and Covalent Bonding

The internuclear distance influences a covalent bond in the same way as it does an ionic bond, since both arise from the balance of attractive and repulsive forces between component atoms. In covalent bonds, we encounter pairs of electrons. The bonding pair is shared between atoms in a compound. The lone pair is not involved in the bonding but belongs solely to one atom. In a coordinate covalent bond, both electrons in the bond come from the same atom.

The Lewis electron structures for molecules and complex ions can be constructed using the following six steps:

1. Arrange the atoms to show which are connected to which.
2. Determine the total number of valence electrons in the molecule or ion.
3. Place a bonding pair of electrons between each pair of adjacent atoms to give a single bond.
4. Add enough electrons to each atom to give all the atoms an octet (or a pair for hydrogen).
5. If any electrons are left over, place them on the central atom.
6. If the central atom does not have an octet, use lone pairs from terminal atoms to form multiple bonds to the central atom.

The use of Lewis dot structures is a good way to visualize the stoichiometry of families of elements, in that they show how many bonds are needed to complete an octet. For Group 17, one bond is needed; for Group 16, two bonds; for Group 15, three bonds; for Group 14, four bonds. In some cases, more than one structure can be drawn for a compound. When this occurs, the structure that yields the lowest formal charge is most likely to be the most stable structure. The formal charge is the difference between the number of valence electrons in a free atom and the number assigned to it in the Lewis structure. When it is possible to write more than one equivalent structure, these are called *resonance structures,* and the actual structure of the compound is the average of the resonances.

8.6. Exceptions to the Octet Rule

There are three exceptions to the octet rule: (1) molecules with an odd number of electrons, (2) molecules containing an atom that has more than eight electrons, and (3) molecules containing an atom with fewer than eight electrons. Molecules with an odd number of electrons do not have all of their electrons paired. Molecules that use *d* orbital electrons to bond can have more than eight valence electrons and are called *expanded-valence molecules.* An electron-deficient molecule contains an atom that has fewer than eight valence electrons; such an atom tends to complete the octet by "borrowing" a lone pair from an adjacent atom.

8.7. Lewis Acids and Bases

In Chapter 4, you learned the definitions of Brønsted–Lowry acids and bases. The Lewis definitions build on those to include other substances. A Lewis acid is a species that can accept a pair of electrons. A Lewis base is a species than can donate a pair of electrons. A coordinate covalent bond forms between Lewis acids and Lewis bases.

8.8. Properties of Covalent Bonds

Covalent bonds have certain distinct properties. The bond order is the number of electron pairs that hold adjacent atoms together. Single bonds have a bond order of one; double bonds have a bond order of two; and triple bonds have a bond order of three. The bond length increases as the bond order increases, as does the bond energy. Bond energies can be used to estimate the enthalpy change for a chemical reaction by subtracting the sum of the bond energies of bonds formed from the sum of the bond energies of bonds broken. If the bonds that are formed are stronger than those that are broken, the reaction is exothermic.

8.9. Polar Covalent Bonds

In between the extremes of ionic bonds and covalent bonds are the polar covalent bonds, where the electrons are not shared equally between the atoms. The polarity of a bond is determined by the electronegativities of the bonded atoms. The polarity and the ionic character both increase with an increasing difference in electronegativities. When the charge distribution is not symmetrical, it produces a dipole moment, which is defined as the product of the partial charge on the bonded atoms and the distance between those partial charges.

Self-Test

1. Which of the following is not a common feature of chemical bonding?
    A. Compounds are more stable than isolated atoms.
    B. Compounds are more volatile than isolated atoms.
    C. A characteristic energy is needed to break the bonds of a compound.
    D. Each chemical bond has a characteristic bond distance.

2. What is the electrostatic energy of one mole of HgI_2 if the internuclear distance is 255.3 pm?
    A. -1.09×10^6 J/mol
    B. 1.09×10^6 J/mol
    C. -1.8×10^{-18} J/ion pair
    D. -545 kJ/mol

3. Lattice energies are highest for substances that contain _____
    A. small ions with low charges.
    B. large ions with low charges.
    C. small ions with high charges.
    D. large ions with high charges.

4. Which of the following is not affected by lattice energy?

  A. Hardness

  B. Dipole moment

  C. Solubility

  D. Melting point

5. The most important factor in determining the stability of an ionic compound is the _____

  A. overall charge.

  B. dipole moment.

  C. bond length.

  D. lattice energy.

6. Calculate the lattice energy of one mole LiCl given an internuclear distance of 251 pm.

  A. −554 kJ/mol

  B. 554 kJ/mol

  C. 1108 kJ/mol

  D. −1108 kJ/mol

7. What is the correct Lewis dot structure for Li?

  A. Li·   B. ·Li⫶

  C. ·Li·   D. ⫶L̈i·

8. What is the correct Lewis dot structure for carbon dioxide?

  A. :Ö:Ö:C̈:   B. :C̈:Ö:C̈:

  C. Ö::C::Ö   D. :Ö:C̈:Ö:

9. What is the correct Lewis structure for HNO?

  A. H:N̈::Ö: B. :H:N̈::Ö:

  C. H:Ö:N̈:   D. :Ḧ:Ö:N̈:

10. The electron pair being shared between two atoms is the _____

  A. lone pair.

  B. valence pair.

  C. bonding pair.

  D. coordinate covalent pair.

11. What is the formal charge on the carbon in CH_2Cl_2?

  A. 4

  B. −4

  C. 2

  D. 0

12. The actual structure of a molecule with multiple resonance structures is _____
    A. the structure that gives the lowest formal charge.
    B. the structure with the lowest overall bond order.
    C. the average of all the resonance structures.
    D. the structure with the highest electronegativity.

13. If an atom can accommodate more than eight valance electrons, it likely has a(n) _____
    A. high lattice energy.
    B. electron-deficient valence.
    C. expanded electron valence.
    D. odd number of electrons.

14. A Lewis acid _____
    A. donates an electron pair.
    B. accepts an electron pair.
    C. donates a proton.
    D. accepts a proton.

15. A Lewis base _____
    A. donates an electron pair.
    B. accepts an electron pair.
    C. donates a proton.
    D. accepts a proton.

16. Which of the following would have the strongest bond?
    A. Be — Be
    B. Mg — Mg
    C. Ca — Ca
    D. Sr — Sr

17. Which of the following would have the weakest bond?
    A. Cl — Cl
    B. Br — Br
    C. F — F
    D. I — I

18. Given these bond energies (all in kJ/mol)

—C—H = 411; O=O = 494; C=O = 799; O—H = 459—

estimate the enthalpy change for the following reaction:

$$CH_4 + 2\,O_2 \rightarrow CO_2 + 2\,H_2O$$

A. -353 kJ/mol

B. -318 kJ/mol

C. -802 kJ/mol

D. -837 kJ/mol

19. The largest factor contributing to the increase in bond polarity is _____

A. bond order.

B. electronegativity.

C. bond length.

D. dipole moment.

20. What is the percent ionic character of CO, given it has a dipole moment of 0.110 D and a bond length of 112.8 pm?

A. 98%

B. 20%

C. 80%

D. 2%

Answers: 1. B; 2. A; 3. C; 4. B; 5. D; 6. B; 7. A; 8. C; 9. A; 10. C; 11. D; 12. C; 13. C; 14. B; 15. A; 16. A; 17. D; 18. C; 19. B; 20. D

CHAPTER 9

Structure and Bonding II: Molecular Geometry and Models of Covalent Bonding

---

Key Words

antibonding molecular orbital
atomic orbitals
bond angles
bonded atoms
bonding molecular orbital
bond order
delocalized
heteronuclear diatomic molecules
homonuclear diatomic molecule
hybrid atomic orbitals
hybridization

linear
linear combinations of atomic orbitals
localized electron-pair bond
molecular geometry
molecular orbital
molecular orbital theory
nonbonding molecular orbital
octahedral
orbital energy
orbitals
pi orbital
pi star orbital

promotion
sigma orbital
sigma star orbital
square planar
tetrahedral
trigonal bipyramidal
trigonal planar
trigonal pyramidal
valence bond theory
valence-shell electron-pair repulsion model (VSEPR)
vectors
wave functions

By the end of this chapter, you should be able to:

- Predict molecular geometry from the VSEPR model
- Predict the dipole moment of a molecule
- Use valence bond theory to describe simple molecules
- Predict bond order from molecular orbital theory
- Use molecular orbitals to explain resonance structures

Chapter Overview

In the previous chapter, we began to discuss the bonding and structure that is possible in molecules. We talked about constructing Lewis dot structures in order to predict the number of bonds around a molecule. In this chapter, we will expand on that and discuss more complex theories of molecular modeling, which will allow us to predict the shape that a molecule will have and predict other factors such as stability.

9.1. Predicting the Geometry of Molecules and Polyatomic Ions

A molecular model known as the *valence-shell electron-pair repulsion model,* or *VSEPR model* for short, will allow us to predict the structure that a molecule is most likely to take. The assumption made in constructing this model is that electron pairs occupy space, and the most likely model is the one that will maximize the distance between electron pairs, thereby minimizing the repulsive forces between them. While the VSEPR model tells us nothing about the bonding of a molecule, it will predict the three-dimensional arrangement and molecular geometry of a molecule. For a molecule with a general formula of AB_n, these are the possible shapes that VSEPR predicts:

1. AB_2, linear or bent
2. AB_3, trigonal planar, trigonal pyramidal, or T-shaped
3. AB_4, tetrahedral or square planar
4. AB_5, trigonal bipyramidal
5. AB_6, octahedral

9.2. Localized Bonding and Hybrid Atomic Orbitals

Valence bond theory is a localized bonding model that works on the assumption that covalent bonds are formed when atomic orbitals overlap and that the bond's strength is proportional to the amount of overlap. Atoms will maximize the overlap with adjacent atoms through a process known as *hybridization*. This occurs when an electron is promoted from a filled ns^2 subshell to a vacant np or $(n-1)d$ valence orbital. The resulting orbitals will then combine to give a new set of equivalent orbitals that have the correct orientation to form bonds. An sp hybrid orbital is formed from the combination of an ns and an np orbital. Three sp^2 orbitals are formed when an ns orbital combines with two np orbitals. Four sp^3 orbitals are formed when an ns orbital combines with three np orbitals. When hybridization involves $(n-1)d$ orbitals, it now becomes possible to have more than eight electrons around the central atom. One $(n-1)d$ orbital will yield five sp^3d orbitals, while two $(n-1)d$ orbitals will yield six sp^3d^2 orbitals.

9.3. Delocalized Bonding and Molecular Orbitals

Molecular orbital theory is a delocalized bonding model that uses the mathematical sums and differences of the wave functions that describe overlapping atomic orbitals. These are called *linear combinations of atomic orbitals* and extend over all the atoms in a molecule or ion. When atomic orbitals combine in such a way that the wave functions are constructively reinforced, these are called *bonding molecular orbitals,* which have lower energy than the parent orbitals. When atomic orbitals combine in such a way that the wave functions experience destructive interference with one another, these are called *antibonding molecular orbitals,* which have higher energy than the parent orbitals. Orbitals that have no influence on the bonding in a molecule or ion are *nonbonding molecular orbitals,* with the same energy as the parent orbitals. There are four geometries of orbitals that arise from this theory: (1) the sigma (σ) orbital is a bonding orbital that is symmetrical about the internuclear axis; (2) the sigma star (σ^*) orbital is an antibonding orbital that is also symmetrical about the internuclear axis; (3) the pi (π) orbital is a bonding orbital with a nodal plane containing the nuclei and electron density along both sides of the plane; and (4) the pi star (π^*) orbital has the same shape and is an antibonding orbital.

An energy level diagram can be constructed from this molecular orbital theory and used to calculate the bond order and to understand the electronic structure of many diatomic molecules. The number of molecular orbitals must be the same as the number of atomic orbitals that interact, and they must be filled in from the bottom up, always following the Paul exclusion principle. When constructing an energy level diagram for a heteronuclear diatomic molecule, the diagram must be skewed toward the more electronegative of the two atoms.

9.4. Combining the Valence Bond and Molecular Orbital Approaches

In order to describe more complex molecules, you must use an approach that combines both the valence bond theory and the molecular orbital theory. This combination uses hybrid atomic orbitals to describe the sigma bonding and molecular orbitals to describe the pi bonding. Unhybridized np orbitals on bonded atoms interact to produce bonding, antibonding, and nonbonding combinations. The resulting energy diagram can be filled in and used to describe bonding that previously required the use of resonance structures.

Self-Test

1. What is the most likely shape for a molecule of $AlCl_3$?
   A. Trigonal pyramidal
   B. Trigonal planar
   C. T-shaped
   D. Tetrahedral

2. What is the most likely shape for a molecule of CCl_4?
   A. Trigonal planar
   B. Trigonal pyramidal
   C. Tetrahedral
   D. Square planar

3. What is the most likely geometry for a molecule having the general formula AB_5?

    A. Octahedral

    B. Trigonal bipyramidal

    C. Tetrahedral

    D. Square planar

4. The molecular geometry is the same as the electron pair geometry when _____

    A. the surrounding atoms have no lone pairs.

    B. the central atom has no lone pairs.

    C. the lone pairs are far enough away to minimize repulsions.

    D. there is only one lone pair in the molecule.

5. What would be the most likely shape of a molecule of NH_3?

    A. Bent

    B. Trigonal planar

    C. Trigonal pyramidal

    D. Tetrahedral

6. The combination of one ns and one np orbital results in a hydrid orbital of _____

    A. sp.

    B. sp^2.

    C. sp^3.

    D. sp^3d.

7. The combination of one ns and all three np orbitals results in hydrids of _____

    A. sp.

    B. sp^2.

    C. sp^3.

    D. sp^3d.

8. What is the hybridization of the boron atom in BF_3?

    A. sp

    B. sp^2

    C. sp^3

    D. sp^3d

9. What is the hybridization of the nitrogen atom in NH_3?

    A. sp

    B. sp^2

    C. sp^3

    D. sp^3d

10. What is the hybridization of the phosphorous atom in PCl_5?

    A. sp

    B. sp^2

    C. sp^3

    D. sp^3d

11. A bonding molecular orbital has energy that is _____ than the parent atom.

    A. higher

    B. lower

    C. no different

12. An antibonding molecular orbital has energy that is _____ than the parent atom.

    A. higher

    B. lower

    C. no different

13. What is the valence electron configuration of N_2?

    A. $(2s\sigma)^2(2s\sigma^*)^2(2p\sigma)^2(2p\sigma^*)^2(2p\pi_x)^2$

    B. $(2s\sigma)^2(2s\sigma^*)^2(2p\sigma)^2(2p\pi_x)^2(2p\pi_x^*)^2$

    C. $(2s\sigma)^2(2s\sigma^*)^2(2p\sigma)^2(2p\pi_x)^2(2p\pi_y)^2$

    D. $(2s\sigma)^2(2p\sigma)^2(2p\sigma^*)^2(2p\pi_x)^2(2p\pi_y)^2$

14. What is the bond order of N_2?

    A. 0

    B. 1

    C. 2

    D. 3

15. What is the valence electron configuration of F_2?

    A. $(2s\sigma)^2(2s\sigma^*)^2(2p\sigma)^2(2p\pi_x)^2(2p\pi_y)^2(2p\pi_x^*)^2(2p\pi_y^*)^2$

    B. $(2s\sigma)^2(2s\sigma^*)^2(2p\sigma)^2(2p\sigma^*)^2(2p\pi_x)^2(2p\pi_x^*)^2(2p\pi_y)^2$

    C. $(2s\sigma)^2(2s\sigma^*)^2(2p\sigma)^2(2p\pi_x)^2(2p\pi_y)^2(2p\pi_z)^2(2p\pi_x^*)^2$

    D. $(2s\sigma)^2(2p\sigma)^2(2p\sigma^*)^2(2p\pi_x)^2(2p\pi_y)^2(2p\pi_x^*)^2(2p\pi_y^*)^2$

16. What is the bond order of F_2?

    A. 0

    B. 1

    C. 2

    D. 3

17. What is the valence electron configuration of O_2?

    A. $(2s\sigma)^2(2s\sigma^*)^2(2p\sigma)^2(2p\sigma^*)^2(2p\pi_x)^2(2p\pi_x{}^*)^1(2p\pi_y)^1$

    B. $(2s\sigma)^2(2s\sigma^*)^2(2p\sigma)^2(2p\pi_x)^2(2p\pi_y)^2(2p\pi_x{}^*)^1(2p\pi_y{}^*)^1$

    C. $(2s\sigma)^2(2s\sigma^*)^2(2p\sigma)^2(2p\pi_x)^2(2p\pi_y)^2(2p\pi_z)^1(2p\pi_x{}^*)^1$

    D. $(2s\sigma)^2(2p\sigma)^2(2p\sigma^*)^2(2p\pi_x)^2(2p\pi_y)^2(2p\pi_x{}^*)^1(2p\pi_y{}^*)^1$

18. What is the bond order of O_2?

    A. 0

    B. 1

    C. 2

    D. 3

19. As the number of interacting atomic orbitals increases, the energy separation between the resulting molecular orbitals _____

    A. increases.

    B. decreases.

    C. stays the same.

20. To describe the bonding in complex molecules, it is best to use _____

    A. atomic orbitals.

    B. hybrid atomic orbitals.

    C. a combination of approaches.

    D. molecular orbitals.

Answers: 1. B; 2. C; 3. B; 4. B; 5. C; 6. A; 7. C; 8. B; 9. C; 10. D; 11. B; 12. A; 13. C; 14. D; 15. A; 16. B; 17. B; 18. C; 19. B; 20. C

CHAPTER 10

Gases

Key Words

absolute zero	effusion	partial pressure
atmosphere	gas constant	pascal
atmospheric pressure	Graham's law	pressure
Avogadro's law	ideal gas law	root mean square speed
barometer	kinetic molecular theory	standard molar volume
Boltzmann distribution	of gases	standard temperature and
Boyle's law	liquefaction	pressure
cryogenic liquid	manometers	torr
Dalton's law	millimeter of mercury	van der Waals equation
diffusion	mole fraction	vapor pressure

By the end of this chapter, you should be able to:

- Describe the characteristics of a gas
- Understand the interconnectedness of the pressure, temperature, amount, and volume of a gas
- Describe the behavior of a gas through the ideal gas law
- Calculate the partial pressures of gases in a mixture
- Understand the kinetic molecular theory of gases
- Recognize the differences between an ideal gas and a real gas

Chapter Overview

We've spent an extensive amount of time talking about the properties of atoms and how atoms combine to form ions and molecules. We now expand our discussion to talk about aggregates that contain large numbers of atoms, ions, or molecules. We begin with the properties and characteristics of gases.

10.1. Gaseous Elements and Compounds

Matter exists in three states: solid, liquid, and gas. Gases have the ability to change both their shape and their volume, and they have the lowest densities. In terms of the periodic table, the elements that are gases at room temperature are found on the table's ride side. These include the noble gases, which are monatomic gases, and the halogens, which are diatomic gases. There are a great number of compounds that exist as gases at room temperature and pressure as well, and all gases are characterized by extremely weak interactions between the atoms or molecules.

10.2. Gas Pressure

There are four quantities needed for a complete physical description of a gas: temperature, volume, pressure, and amount. *Pressure* is defined as the force per unit area of surface, with a metric unit of a pascal, or one newton per square meter. The pressure exerted by the earth's atmosphere is called *atmospheric pressure* and is measured with a barometer, which is a closed and inverted tube filled with mercury. The height of the mercury column is proportional to the atmospheric pressure and is reported in millimeters of mercury or torr. One atmosphere is defined as the pressure required to support a column of mercury that is 760 mm tall. A manometer is a device used to measure the pressure of a sample of gas.

10.3. Relationships between Pressure, Temperature, Amount, and Volume

Boyle's law states that the volume of a sample of gas is inversely proportional to its pressure. Charles's law states that the volume of a gas is directly proportional to its temperature at constant pressure. Avogadro's law states that the volume of a gas is directly proportional to the number of moles of gas in the sample. If you plot the volume of a gas versus the temperature and extrapolate to zero volume, you reach the temperature known as *absolute zero*.

10.4. The Ideal Gas Law

The relationships between the temperature, pressure, volume, and amount of a gas can all be combined into a single equation known as the ideal gas law, $PV = nRT$, where R is a proportionality constant known as the *gas constant* and has different values depending upon which units are used. This equation describes the behavior of a hypothetical ideal gas, and most real gas samples will not have ideal gas properties. Standard temperature and pressure is defined as 0°C and 1 atm. The standard molar volume of an ideal gas is defined as 22.4 L. The ideal gas law allows you to calculate any variable if the other three are known and to predict the changes in the other variables caused by a change in any one of them. The law can also be used to calculate a gas's density or molar mass.

10.5. Mixtures of Gases

The pressure exerted by each gas in a mixture is called the *partial pressure* and is independent of the pressure exerted by all other gases in the mixture. Dalton's law of partial pressures states that the total pressure of a mixture of gases is the sum of the partial pressures of each gas present in the mixture. The amount of a gas in a mixture may also be described by the mole fraction, the ratio of the number of that gas's moles to the total number of moles of all gases in the mixture. You can calculate the partial pressure of any gas in a mixture from that gas's total pressure and mole fraction.

10.6. Gas Volumes and Stoichiometry

The amount of products and reactants in a reaction involving gases can be calculated from the stoichiometry of the reaction, along with the volume of the gases. The ideal gas law can be useful here, along with the standard molar volume of a gas. Gases generated through an experiment in the laboratory are often collected through a process called *displacement,* in which the gas displaces a liquid from inside a graduated container. This gas is not pure, and calculations must correct for water vapor that may be contained in the gas.

10.7. The Kinetic Molecular Theory of Gases

The behavior of ideal gases is explained through the kinetic molecular theory of gases. The high amount of molecular motion in a gas leads to collisions between the molecules of the gas and between the molecules and container walls, which gives rise to the pressure of the gas. Gases have a high intermolecular distance, which causes the high compressibility of a gas. All gases have the same average kinetic energy but not the same root mean square speed. A Boltzmann distribution shows the speed and kinetic energy for the particles of a gas, with some falling above and below the average for the gas. *Diffusion* is the process by which gases will gradually mix, with no external agitation, to form a uniform composition. *Effusion* is the process by which a gas will escape a closed container through tiny openings. Graham's law states that the rates of effusion and diffusion are inversely proportional to the square of the molar mass of a gas.

10.8. The Behavior of Real Gases

Real gases do not exhibit ideal gas behavior, though many gases may approximate it over a range of conditions. Gases tend to deviate from idea behavior at higher pressures and lower temperatures. Deviations from the ideal gas law can be described by the van der Waals equation, which includes corrections for the characteristics of real gases. Liquefaction occurs when a gas reaches a low enough temperature to pass into the liquid phase. Liquid gases have many uses in industry and are much easier to transport than gas phase gases.

Self-Test

1. Which of the following is not a characteristic of a gas?
   A. Low density
   B. High compressibility
   C. Strong intermolecular forces
   D. Filling a container completely

2. Gases are most generally found in what area of the periodic table?
   A. Upper left
   B. Upper right
   C. Lower left
   D. Middle

3. How many torr are in 1.15 atm?
   A. 850
   B. 874
   C. 892
   D. 908

4. The SI unit for pressure is the _____
   A. pascal.
   B. torr.
   C. millimeters of mercury.
   D. atmosphere.

5. The statement that the volume of a gas is inversely proportional to its pressure is known as _____
   A. Boyle's law.
   B. Charles's law.
   C. Avogadro's law.
   D. the ideal gas law.

6. The statement that the volume of a gas is directly proportional to its temperature is known as _____
   A. Boyle's law.
   B. Charles's law.
   C. Avogadro's law.
   D. the ideal gas law.

7. What is the volume of 1 mol of an ideal gas under conditions of standard temperature and pressure?

   A. 22.14 L

   B. 22.26 L

   C. 22.39 L

   D. 22.41 L

8. What is the volume of 2 moles of an ideal gas at 278 K and 0.97 atm?

   A. 52.25 L

   B. 55.35 L

   C. 50.42 L

   D. 58.19 L

9. What is the volume of 1.5 moles of an ideal gas at 273.15 K and 0.75 atm?

   A. 44.28 L

   B. 44.52 L

   C. 44.78 L

   D. 44.82 L

10. How many moles of an ideal gas are contained in a volume of 29 L at standard temperature and pressure?

    A. 1.15

    B. 1.29

    C. 1.34

    D. 1.41

11. The statement that total pressure of a mixture of gases is the sum of the partial pressures of the component gases is known as _____

    A. Charles's law.

    B. Avogadro's law.

    C. Dalton's law.

    D. Graham's law.

12. A mixture of hydrogen and oxygen gases has a total pressure of 9 atm. The pressure of the hydrogen is 5 atm. What is the volume of the oxygen gas at 298 K if there are 4 moles present?

    A. 22.41 L

    B. 22.45 L

    C. 22.49 L

    D. 22.52 L

13. What is the mole fraction of oxygen in Question 12?

    A. 23%

    B. 34%

    C. 44%

    D. 56%

14. Gas evolved by reactions in the laboratory is generally collected through a process called _____

    A. sublimation.

    B. displacement.

    C. liquefaction.

    D. effusion.

15. Gases collected in the laboratory usually have to be corrected for _____

    A. temperature.

    B. pressure.

    C. vapor pressure.

    D. density.

16. The mathematical equation that explains the rates of effusion and diffusion is known as _____

    A. Charles's law.

    B. Avogadro's law.

    C. Dalton's law.

    D. Graham's law.

17. Gas A has a molar mass of 17 g/mol, and gas b has a molar mass of 38.25 g/mol. What is the ratio of diffusion of these two gases?

    A. 1.5

    B. 2.25

    C. 5.06

    D. 4.4

18. What is the ratio of effusion of the two gases from Question 17?

    A. 1.5

    B. 2.25

    C. 5.06

    D. 4.4

19. Real gases tend to exhibit ideal behavior under which of the following conditions?

    A. High temperature and high pressure

    B. High temperature and low pressure

    C. Low temperature and high pressure

    D. Low temperature and low pressure

20. The condensation of gases into a liquid is known as _____

    A. sublimation.

    B. condensation.

    C. liquefaction.

    D. cryogenation.

Answers: 1. C; 2. B; 3. B; 4. A; 5. A; 6. B; 7. D; 8. C; 9. D; 10. B; 11. C; 12. B; 13. C; 14. B; 15. C; 16. D; 17. A; 18. A; 19. B; 20. C

CHAPTER 11

Liquids

Key Words

adhesive forces
anisotropic
capillary action
cholesteric
cohesive forces
condensation
cooling curve
critical point
critical pressure
critical temperature
dipole-dipole interactions
dynamic equilibrium
enthalpy of fusion
enthalpy of sublimation
equilibrium vapor
  pressure

evaporation
heating curve
hydrogen bonds
induced dipole
instantaneous dipole
  moment
ionic liquid
isotropic
liquid crystals
London dispersion forces
meniscus
molten salt
nematic
nonvolatile liquids
normal boiling point
phase changes

phase diagram
polarizability
smectic
sublimation
supercooled liquid
supercritical fluid
superheated liquid
surface tension
surfactants
triple point
van der Waals forces
vaporization
vapor pressure
viscosity
volatile liquids

By the end of this chapter, you should be able to:

- Understand the kinetic molecular description of liquids
- Describe the intermolecular forces in liquids
- Describe the properties of liquids
- Describe how the temperature of a liquid influences its vapor pressure
- Read a phase diagram
- Understand phase changes and calculate the energies that go along with them
- Understand critical temperatures and pressures of liquids

Chapter Overview

In the last chapter, we focused on the properties of gases. In this chapter, we continue our discussion of phases of matter by talking about liquids. We discuss the intermolecular forces at work in liquids and talk about some of their unique properties. We also cover phase changes.

11.1. The Kinetic Molecular Description of Liquids

In the last chapter, we used the kinetic molecular theory to describe the behavior of gases. This theory can also be used to describe the characteristics of liquids. The theory must be modified to consider the volumes of the particles and the intermolecular forces of the liquids, but otherwise it is a good description of liquid behavior. This theory explains why liquids take on the shape of their containers. It also explains the density, molecular order, low compressibility, diffusion, and thermal expansion of liquids.

11.2. Intermolecular Forces

Molecules in liquids are held together by three types of fairly weak intermolecular interactions: dipole-dipole interactions, London dispersion forces, and hydrogen bonds. Dipole-dipole interactions arise from the electrostatic interactions of the positive and negative ends of molecules that have a permanent dipole moment. The strength of these interactions depends upon the magnitude of the dipole moments and on the distance between them. London dispersion forces arise from fluctuations of electron charge distribution, which leads to a brief formation of an induced dipole on adjacent atoms. The strength of these interactions also depends upon the distance between the atoms. Hydrogen bonds arise from dipole-dipole interactions between hydrogen and highly electronegative atoms like oxygen, nitrogen, and fluorine. This results in a partial positive charge on the hydrogen and a partial negative change on the electronegative atom.

11.3. Unique Properties of Liquids

The amount of energy required to raise the surface area of a liquid by a given amount is known as the *surface tension,* and it is dependent upon the strength of the intermolecular interactions between molecules. Surfactants are molecules that work to reduce the surface tension of liquids. These types of molecules are quite useful as soaps and detergents. *Capillary action* is the process that allows liquid to rise up the sides of a narrow tube. This occurs when the cohesive forces that attract the molecules to one another are weaker than the adhesive forces that attract the molecules to the surface of the tube. These forces also determine the shape of the meniscus, the upper surface of the liquid in the tube. A liquid's viscosity is its resistance to flow, and this is also determined by the strength of the intermolecular interactions.

11.4. Vapor Pressure

There is always some fraction of molecules in a liquid that have enough kinetic energy to escape the surface of the liquid and enter the gas phase. This process of vaporization generates a vapor pressure above the liquid. Molecules that have vaporized can reenter the liquid phase by colliding with the liquid's surface and condensing. A dynamic equilibrium is established when the number of molecules entering and leaving the liquid phase per unit time is equal. This gives a liquid its characteristic equilibrium vapor pressure. Volatile liquids are those with high vapor pressures, while nonvolatile liquids are those with low vapor pressures.

11.5. Changes of State

Phase changes, or changes of state, occur when a molecule transitions between solid, liquid, or gas phases. Each of these changes have their own characteristic energy. Changes that move from a less ordered state to a more ordered state give off energy to their surroundings and are exothermic. Changes that move from a more ordered state to a less ordered state take in energy from their surroundings and are endothermic. The conversion of a solid to a liquid is known as *fusion* and is accompanied by the enthalpy of fusion. The conversion of a liquid to a gas is known as *vaporization* and is accompanied by the enthalpy of vaporization. The conversion of a solid to a gas is known as *sublimation* and is accompanied by the enthalpy of sublimation. The enthalpy of sublimation is the sum of the enthalpies of fusion and vaporization. The heating curve of a substance is constructed by plotting the temperature versus the heat added. The cooling curve of a substance is constructed by plotting the temperature versus the heat released. A superheated liquid is an unstable liquid at a temperature where the substance should be a gas. A supercooled liquid is a metastable liquid at a temperature where the substance should be a solid. Supercooled liquids are likely to crystallize.

11.6. Critical Temperature and Pressure

The critical temperature is the temperature above which the substance cannot exist as a liquid, regardless of the pressure applied. The minimum pressure needed to liquefy a substance at its critical temperature is known as the *critical pressure*. The combination of the critical temperature and critical pressure is known as the substance's *critical point*. A substance may exist above its critical temperature as a supercritical fluid, which has the gaseous characteristic of filling its container but has a liquidlike density.

11.7. Phase Diagrams

A phase diagram is a plot of temperature versus pressure and shows the various physical states of a substance. These diagrams contain distinct regions for the solid, liquid, and gas phases. The solid and liquid phases are separated by the melting curve of a substance, while the liquid and gas phases are separated by the vapor pressure curve, ending in the critical point. Along these lines, both phases are present in equilibrium. The melting and vapor curves intersect at the triple point, which is the only temperature and pressure at which all three phases are present.

11.8. Liquid Crystals

Liquid crystals are a class of substances that have properties between those of a crystalline solid and a normal liquid; they are typically long, rigid structures that interact very strongly with one another. Liquid crystals display the unique property of having anisotropic structures, which means they exhibit different properties when viewed from different directions. In the nematic phase, the long axes of the molecules are aligned. In the smectic phase, the long axes are parallel, resulting in the molecules lying in planes. In the cholesteric phase, the molecules are arranged in planes that rotate layers, resulting in a helical structure.

Self-Test

1. Which of the following is not a characteristic of a liquid?
   A. Low compressibility
   B. Short-range order
   C. Low density
   D. Lack of thermal expansion

2. Which of the following would you expect to have the highest boiling point?
   A. Isobutane
   B. Water
   C. Acetonitrile
   D. Ethane

3. Which of the following would you expect to have the highest boiling point?
   A. Methane
   B. Propane
   C. Pentane
   D. Hexane

4. Which of the following is not a weak intermolecular force responsible for the characteristics of a liquid?
   A. Dipole–dipole interactions
   B. Covalent interactions
   C. London dispersion forces
   D. Hydrogen bonds

5. Mercury in a capillary tube has a _____ meniscus.
   A. concave
   B. convex

6. Strong adhesive forces result in _____
   A. surface tension.
   B. capillary action.
   C. viscosity.
   D. high polarity.

7. Which of the following would you expect to have the highest surface tension?
   A. Diethyl ether
   B. Acetone
   C. Ethanol
   D. Ethylene glycol

8. Which of the following would you expect to be the most volatile?

   A. Octanol

   B. Hexane

   C. Acetone

   D. Diethyl ether

9. The process of escaping from the surface of a liquid to enter the gas phase is known as _____

   A. condensation.

   B. sublimation.

   C. vaporization.

   D. fusion.

10. Nonvolatile liquids have _____

    A. high vapor pressures and high boiling points.

    B. high vapor pressures and low boiling points.

    C. low vapor pressures and high boiling points.

    D. low vapor pressures and low boiling points.

11. The process of passing from the solid phase to the gas phase is known as _____

    A. condensation.

    B. sublimation.

    C. vaporization.

    D. fusion.

12. A change from the liquid state to the solid state would be _____

    A. endothermic.

    B. exothermic.

13. The process of passing from the solid phase to the liquid phase is known as _____

    A. condensation.

    B. sublimation.

    C. vaporization.

    D. fusion.

14. A substance cannot form a liquid above its _____

    A. critical pressure.

    B. critical temperature.

    C. critical point.

    D. triple point.

15. Which of the following would you expect to have the highest critical pressure?
    A. Carbon dioxide
    B. Ethanol
    C. Water
    D. Mercury

16. Which of the following is not a characteristic of substances with high critical temperatures?
    A. High density
    B. High boiling points
    C. Nonvolatile
    D. Strong intermolecular forces

17. The solid phase is favored under the conditions of _____
    A. high temperature and high pressure.
    B. high temperature and low pressure.
    C. low temperature and high pressure.
    D. low temperature and low pressure.

18. All phases are present at the _____
    A. critical point.
    B. melting point.
    C. boiling point.
    D. triple point.

19. Which of the following is not a characteristic of a liquid crystal?
    A. Rigid molecules
    B. Nonpolar groups
    C. Ordered arrangement
    D. Ability to flow

20. Having the axes of the molecules in a liquid crystal aligned parallel results in the _____
    A. anisotropic phase.
    B. nematic phase.
    C. smectic phase.
    D. cholesteric phase.

Answers: 1. C; 2. C; 3. D; 4. B; 5. B; 6. B; 7. D; 8. D; 9. C; 10. C; 11. B; 12. B; 13. D; 14. B; 15. D; 16. A; 17. C; 18. D; 19. B; 20. C

CHAPTER 12

Solids

Key Words

alloy
amorphous solid
band gap
band theory
bandwidth
BCS theory
body-centered cubic
Bragg equation
ceramic
cesium chloride structure
composite material
conduction band
covalent solids
crystal lattice
crystalline solid
crystals
cubic close-packed structure
doping
edge dislocation
electrical insulators
energy band
enzymes
face-centered cubic
fibers

Frenkel defect
grain boundary
hexagonal close-packed structure
high-temperature superconductors
intermetallic compound
interstitial alloy
interstitial impurity
ionic solids
line defect
Meissner effect
metallic solids
metal-matrix composite
molecular solids
monomers
nonstoichiometric compounds
n-type semiconductor
overlapping bands
peptide bonds
peptides
perovskite structure
pinning
plane defect

plastic
point defect
polymer-matrix composite
polymers
proteins
p-semiconductor
pyrolysis
Schottky defect
semiconductors
simple cubic
sodium chloride structure
sol-gel process
solid electrolytes
substitutional alloy
substitutional impurity
superalloys
superconducting transition temperature
superconductivity
superconductors
unit cell
vacancy
work hardening
X-ray diffraction
zinc blende structure

By the end of the chapter, you should be able to:

- Identify the unit cell of a crystalline solid
- Calculate the density of a solid from its unit cell
- Understand the origin and nature of defects in crystals
- Understand how X-rays are diffracted by crystalline solids
- Understand how bonding affects the properties of solids

- Use band theory to describe the electrical properties of a solid
- Describe the properties of superconductors
- Understand the differences between synthetic and biological polymers
- Describe the properties of some contemporary materials

Chapter Overview

In the last two chapters, we looked at the structures and characteristics of gases and liquids. We now complete this portion of the discussion by talking about the properties of solids. Solids are different from liquids and gases in that they are locked into shape and volume by rigid structures. Many solids possess long-range, 3-D patterns that lead to a wide range of properties.

12.1. Crystalline and Amorphous Solids

Solids have the unique property of long-range, three-dimensional arrangements of atoms, ions, or molecules that lock them into position. The components of a solid can be arranged in either a regular repeating three-dimensional array called a *crystal lattice* (which results in a crystalline solid), or it can move more or less randomly to produce an amorphous solid. Crystalline solids possess well-defined edges and faces. They also diffract X-rays and tend to have sharp melting points. Amorphous solids have irregular or curved surfaces. They do not produce well-resolved X-ray diffraction patterns and tend to have a wider melting point.

12.2. The Arrangement of Atoms in Crystalline Solids

The smallest repeating unit of a crystal lattice is known as the *unit cell*. There are several different kinds of unit cells. The simple cubic cell contains eight atoms, molecules, or ions that hold positions at the corners of a cube. A body-centered cubic cell adds one more component in the cube's center. A face-centered cubic cell has components at the corners of the cube and additional components in the center of each face of the cube. The hexagonal close-packed structure has a repeating arrangement of two components. The cubic close-packed structure has a repeating arrangement of three components and resembles a face-centered cubic lattice.

12.3. Structures of Simple Binary Compounds

The packing arrangement of the largest species, the anions, present in a binary compound determines its structure, with the smaller species, the cations, occupying holes in the anion lattice. The simplest arrangement is the simple cubic lattice, which has one hole in the center of the unit cell. Placing a cation in that hole results in the cesium chloride structure, which has a 1:1 cation:anion ratio and a coordination number of eight. A face-centered cubic array contains both octahedral and tetrahedral holes. If the octahedral holes are filled, the result is a sodium chloride structure, with a 1:1 ratio of cations to anions and a coordination number of six. If half of the tetrahedral holes are occupied, the result is a zinc blende structure, with a 1:1 ratio of cations to anions and a coordination number of four. X-ray diffraction can be used to determine the structures of crystalline solids, since the wavelength of X-ray radiation is comparable to the interatomic distances of most solids. The X-rays diffracted from different planes in a solid can reinforce each other if they are in phase and if the extra distance they travel is equal to an integral number of wavelengths. This phenomena is described by the Bragg equation.

12.4. Defects in Crystals

Crystalline solids are not perfect and in fact have a high number of defects. Defects can be classified by how much of the lattice they affect. A point defect occurs at a single point in the lattice, while a line defect occurs in a row of lattice points. A plane defect occurs when a defect affects an entire plane of atoms. Defects may also be characterized by their type. An atom missing from a site in a crystal is a *vacancy*. An impurity atom can occupy a normal lattice site in a substitutional impurity, or a hole in the lattice between atoms in an interstitial impurity. In a defect known as an *edge dislocation*, an extra plane of atoms is inserted into part of the crystal lattice. Just as there are multiple types of defects, there are multiple ways of controlling defects. Additional defects can be introduced into the lattice so that the motion of one will restrict the motion of another, in a process known as *pinning*. The motion of a defect tends to stop at the grain boundaries, so controlling the size of the grains can control the defects. Sometimes defects are introduced into the lattice to harden metals, in a process called *work hardening*. A Schottky defect occurs when there is a coupled vacancy, one cation and one anion, which results in maintained electrical neutrality. A Frenkel defect occurs when an ion occupies an incorrect site in the lattice and usually results in the compound having high conductivity.

12.5. Correlation between Bonding and the Properties of Solids

There are four major types of solids: ionic, molecular, covalent, and metallic. An ionic solid consists of cations and anions that are held together by electrostatic forces. Ionic solids tend to be hard and have high melting points. Molecular solids are held together by weaker forces such as hydrogen bonding, dipole-dipole interactions, and London dispersion forces. They tend to be softer than ionic solids and have lower melting points. Covalent solids are made up of covalently bonded two- or three-dimensional arrays of atoms. These are strong solids that tend to be very hard and have high melting points. Metallic solids consist of metal atoms and have the unusual properties of high thermal and electrical conductivity; they are also malleable and ductile and exhibit luster. An alloy is a mixture of metals that has properties different from those of the separate metals and can be formed by substituting one metal for another in the lattice or by inserting a metal into holes in the lattice or by a combination of methods.

12.6. Bonding in Metals and Semiconductors

Band theory works on the assumption that the valence orbitals of the atoms in a solid interact to generate a set of molecular orbitals that extend throughout the solid, with the continuous set of allowed energy levels constituting an energy band. The bandwidth is the difference between the highest and lowest allowed levels within the band, and the bandgap is the difference between the highest level of the band

and the lowest level of the band above it. Overlapping bands result when the width of adjacent bands is larger than the energy gap between them. If a solid has a filled band with an empty band above it, electrons can possibly be excited into that empty band, from which they can migrate through the crystal and create electrical conductivity. Electrical insulators are poor conductors because their valence bands are full. Semiconductors have electrical properties between those of insulators and metals; these properties can be modified through a process called *doping*, which involves introducing impurities. Adding an element with more valence electrons than the host atom results in an *n*-type semiconductor with increased conductivity. Adding an element with fewer valence electrons than the host atom results in a *p*-type semiconductor, which also exhibits increased conductivity.

12.7. Superconductors

Superconductors are solids that show zero resistance to the flow of electrical currents at low temperatures. The temperature at which this occurs is called the *superconducting transition temperature*. Superconductors also produce a magnetic field, a phenomenon known as the *Meissner effect*. Superconductivity is explained through the BCS theory, which states that electrons are able to travel through solids with no resistance because they couple to form pairs of electrons.

12.8. Polymeric Solids

Polymers are giant molecules that consist of long chains of repeating units known as *monomers*. Small biological polymers formed from amino acids are called *peptides*, while larger ones are called *proteins*. A particle that is more than 100 times longer than it is wide is known as a *fiber*.

12.9. Contemporary Materials

Ceramics are strong but brittle nonmetallic, inorganic substances with high melting points. There are two major classes of modern ceramics—the ceramic oxides and the nonoxide ceramics, which consist of carbides or nitrides. Making ceramics usually involves pressing a powder into a mold and sintering a temperature just below the boiling point. Superalloys are new metals based on cobalt, nickel, or iron that have unusually high temperature stability and resistance to oxidation. Composite materials consist of at least two phases, and there are several types of these. Polymer-matrix composites have fibers embedded in a polymer matrix. Metal-matrix composites have fibers of boron, graphite, or ceramic in a metal matrix. Ceramic-matrix composites use fibers of ceramic to reinforce a ceramic matrix.

Self-Test

1. A regular, repeating three-dimensional structure in a solid is known as a(n) _____
   A. amorphous solid.
   B. crystal lattice.
   C. unit cell.
   D. simple cubic.

2. A solid that aggregates with no particular order is known as a(n) _____
   A. crystalline solid.
   B. crystal lattice.
   C. amorphous solid.
   D. unit cell.

3. The smallest repeating unit in a crystalline solid is the _____
   A. unit cell.
   B. face-centered cubic.
   C. monomer.
   D. array.

4. The unit cell that has components in the corners and in the center of a cube is the _____
   A. simple cubic.
   B. body-centered cubic.
   C. face-centered cubic.
   D. cubic close-packed.

5. An ionic solid has a face-centered cubic structure with cations occupying the octahedral holes. This results in what type of structure?
   A. Cesium chloride structure
   B. Sodium chloride structure
   C. Zinc blende structure
   D. Perovskite structure

6. Which of the following variables is not included in the Bragg equation?
   A. Angle of incidence
   B. Distance between planes of atoms
   C. Frequency of radiation
   D. Wavelength of radiation

7. Which of the following is not a type of defect found in a crystal lattice?

   A. Point defect

   B. Array defect

   C. Plane defect

   D. Line defect

8. A Schottky defect results in _____

   A. electrical positivity.

   B. electrical negativity.

   C. electrical conductivity.

   D. electrical neutrality.

9. An interstitial impurity arises when an atom occupies _____

   A. a normal space in the lattice.

   B. an extra plane in the lattice.

   C. a hole between the atoms in a lattice.

   D. a space occupied by an excited atom in the lattice.

10. A solid held together by London dispersion forces would be an example of a(n) _____

   A. ionic solid.

   B. molecular solid.

   C. covalent solid.

   D. metallic solid.

11. Which of the following is not a property of a metallic solid?

   A. Conducts electricity

   B. Malleable

   C. Brittle

   D. Lustrous

12. A solid with filled valence bands would be a(n) _____

   A. semiconductor.

   B. insulator.

   C. n-type semiconductor.

   D. p-type semiconductor.

13. Which of the following would not be an electrical conductor?

   A. Copper

   B. Iron

   C. Silicon

   D. Glass

14. The energy difference between the highest and lowest energies of a band is the _____
    A. band energy.
    B. band gap.
    C. bandwidth.
    D. band overlap.

15. Conductivity is _____ to resistivity.
    A. proportional
    B. inversely proportional

16. The Meissner effect is the expulsion of a _____ from a superconductor.
    A. magnetic field
    B. pair of electrons
    C. cation
    D. anion

17. The bond between the carbon and nitrogen joining two amino acids is called a(n) _____
    A. protein bond.
    B. chain linkage.
    C. amino bond.
    D. peptide bond.

18. The smallest repeating unit that makes up a polymer is the _____
    A. monopolymer.
    B. copolymer.
    C. monomer.
    D. comer.

19. Which of the following is not included in the major classes of ceramics?
    A. Oxides
    B. Carbides
    C. Phosphides
    D. Nitrides

20. Ceramic-matrix composites are made up of _____
    A. polymer fibers in a ceramic matrix.
    B. ceramic fibers in a ceramic matrix.
    C. ceramic fibers in a metal matrix.
    D. ceramic fibers in a polymer matrix.

Answers: 1. B; 2. C; 3. A; 4. B; 5. B; 6. C; 7. B; 8. D; 9. C; 10. B; 11. C; 12. B; 13. D; 14. C; 15. B; 16. A; 17. D; 18. C; 19. C; 20. B

CHAPTER 13

Solutions

Key Words

aerosols	fractional crystallization	phospholipids
amalgams	freezing point depression	Raoult's law
boiling point elevation	Henry's law	saturated
cell	hydration	seed crystal
cell membrane	hydrophilic	semipermeable membrane
colligative properties	hydrophobic	solubility
colloid	ideal solution	solvation
concentration	ion pairs	supersaturated
crown ethers	mass percentage	suspension
cryptands	micelles	thermal pollution
crystallization	miscible	Tyndall effect
dialysis	molality	unsaturated
dielectric constant	osmosis	van't Hoff factor
emulsions	osmotic pressure	water soluble
entropy	parts per billion	
fat soluble	parts per million	

By the end of this chapter, you should be able to:

- Understand how solution formation is affected by enthalpy and entropy
- Understand how molecular structure influences solubility
- Describe the concentration of a solution in a number of different ways
- Understand the relationships between temperature, pressure, and solubility
- Describe how the concentration of the solute affects the colligative properties of the solution
- Define what constitutes a true solution

Chapter Overview

In the previous three chapters, we've discussed each of the physical states of matter in turn. In doing so, we've restricted ourselves to pure substances. In reality, though, you're much more likely to encounter a mixture, either heterogeneous or homogeneous. A homogeneous mixture is known as a *solution,* and that is what we discuss in this chapter.

13.1. Factors Affecting Solution Formation

When two or more substances are uniformly distributed on a microscopic scale, you have a homogeneous mixture known as a *solution.* A solution is broken up into two parts: the solvent (the substance present in the larger amount) and the solute (the substance present in the lesser amount). Forming a solution is not a chemical process but is a physical one. Substances that mix to form a single phase are said to be *miscible,* while those that mix to form more than one phase are said to be *immiscible.* Solvation occurs when solute particles are surrounded by solvent molecules. When this happens with water, it is known as *hydration.* The energy change that goes with the formation of a solution is the sum of the enthalpies of breaking the bonds between the solute molecules and solvent molecules and the formation of bonds between the solute and solvent. The formation of a solution tends to be exothermic. In addition to the enthalpies of the system, the entropy, or degree of disorder in the system, must also be taken into account. A decrease in order results in an increase in entropy, and this favors dissolution.

13.2. Solubility and Molecular Structure

A substance has a characteristic solubility, which is the maximum amount of solute that can dissolve in a given quantity of solvent. Solubilities are dependent upon the chemical structure of the solute and solvent, as well as the temperature and pressure. A solution is said to be *saturated* when it contains the maximum amount of solute that can dissolve under a given set of conditions. Any amount less than this gives an unsaturated solution. An amount greater than this gives a supersaturated solution, which is unstable and very easy to crystallize. A system is said to be in *dynamic equilibrium* when crystallization and dissolution occur at the same rate. Solutes can be classified as either hydrophilic, meaning water-loving, or hydrophobic, meaning water-fearing. Vitamins that contain hydrophilic structures are water-soluble and only remain in our bodies a short time. Vitamins that contain hydrophobic structures are fat-soluble, which means they can remain in our bodies and accumulate over a longer period of time. Network solids and most metals are insoluble in nearly all solvents, though many metals will dissolve in liquid mercury to form amalgams. The dielectric constant of a solvent is the ability to decrease the electrostatic forces between charged particles, and this determines the solubility of an ionic compound.

13.3. Units of Concentration

A solution's concentration is the amount of solute in a given quantity of solution, and this can be expressed in ways such as molarity, mole fraction, mass percentage, parts per million, parts per billion, or molality. The first three terms were defined in previous chapters and should already be familiar to you. Parts per million is defined as the number of milligrams of solute per kilogram of solution; parts per billion is the number of micrograms of solute per kilogram of solution. The molality of a solution is the number of moles of solute per kilogram of solvent.

13.4. Effects of Temperature and Pressure on Solubility

The solubility of a substance depends on the temperature and pressure of that substance, with solubility for liquids and solids generally increasing with a rise in temperature. A process known as *fractional crystallization* is used to separate the components of a mixture based on their relative solubilities. The solubility of a gas tends to decrease with an increase in temperature. The relationship between the solubility of a gas and its pressure is shown by Henry's law.

13.5. Colligative Properties of Solutions

A colligative property of a solution is something that depends upon only the number of dissolved particles in the solution, not on the chemical identity of the particles. There are several of these, including vapor pressure, boiling point, freezing point, and osmotic pressure. Raoult's law defines the relationship between the vapor pressure of a solution and the mole fraction of the solute in solution; a solution that obeys this law is said to be an *ideal solution*. Like the ideal gas law, this relationship is not often seen in real solutions. Most real solutions will exhibit either positive or negative deviations from the law. The boiling point elevation and freezing point depression of a solution are defined as the differences between the boiling and freezing points of the solution and the pure solvent; these are both determined by the molality of the solute. A solution and a pure solvent can be separated by a semi-permeable membrane, which will not allow the solute to pass through. This creates an osmotic pressure, where the flow of solvent is unequal across the membrane. Osmosis is the net flow of solvent through the membrane from a region of higher concentration into a region of lower concentration. Dialysis is a process that uses a membrane capable of passing small solute molecules in addition to solvent. The colligative properties of a solution can be altered when cations and anions in concentrated or highly charged solutions combine to form ion pairs. The ratio of the apparent number of particles in solution to the number predicted by the stoichiometry of the salt is the van't Hoff factor, and this describes the extent of ion pair formation.

13.6. Aggregate Particles in Aqueous Solution

A heterogeneous mixture of particles of a substance distributed through a second phase is known as a *suspension*. A suspension will separate out over time. A colloid, however, contains smaller particles and will not separate over time. There are three types of colloids: (1) an emulsion is a dispersion of one liquid in another; (2) an aerosol is a dispersion of a liquid or solid in a gas; and (3) a sol is a dispersion of solid particles in another solid or in a liquid. A colloid can be difficult to visually distinguish from a true solution. However, a colloid has the unique characteristic of being able to scatter a beam of light; this is called the *Tyndall effect*. An emulsion is formed by dispersing a hydrophobic liquid in water. In the absence of an emulsion, solutions of detergents in water will form organized aggregates known as *micelles*.

Phospholipids are made up of hydrophobic tails attached to hydrophilic heads. This allows the head of the molecule to interact with water, while the tail interacts with grease or some other form of hydrophobic material.

Self-Test

1. Substances that form a single homogeneous phase in all proportions are _____
   A. solvated.
   B. hydrated.
   C. miscible.
   D. colloidal.

2. A polar solute is most likely to dissolve in a solvent that is _____
   A. slightly polar.
   B. polar.
   C. slightly nonpolar.
   D. nonpolar.

3. A homogeneous mixture is also known as a(n) _____
   A. solution.
   B. emulsion.
   C. colloid.
   D. suspension.

4. A solution that contains the maximum possible amount of solute is said to be _____
   A. unsaturated.
   B. saturated.
   C. supersaturated.
   D. crystallized.

5. Hydrophobic substances tend to be most soluble in _____
   A. water.
   B. polar solvents.
   C. neutral solvents.
   D. nonpolar solvents.

6. Which of the following does not affect the solubility of a substance?
   A. Temperature
   B. Pressure
   C. Atomic radius
   D. Polarity

7. A good measure of the tendency of a solvent to dissolve ionic compounds is the _____

    A. enthalpy.

    B. entropy.

    C. solubility.

    D. dielectric constant.

8. What is the parts-per-billion concentration of a solution with a mass of 25 grams and a solute mass of 7.5 micrograms?

    A. 3

    B. 30

    C. 300

    D. 3000

9. What is the parts-per-million concentration of a solution with a mass of 32.3 grams and a solute mass of 2.1 micrograms?

    A. 0.065

    B. 0.65

    C. 6.5

    D. 65

10. What is the molality of a solution containing 250 grams of NaCl in 1 kilogram of water?

    A. 0.22

    B. 2.2

    C. 4.3

    D. 8.6

11. As the temperature of a system increases, the solubility of a substance _____

    A. increases.

    B. decreases.

    C. stays the same.

12. The relationship between pressure and the solubility of a gas is given by _____

    A. Raoult's law.

    B. Henry's law.

    C. the van't Hoff factor.

    D. the Tyndall effect.

13. Colligative properties depend upon _____

    A. polarity.

    B. pressure.

    C. number of particles.

    D. temperature.

14. The relationship between solution composition and vapor pressure is given by _____

    A. Raoult's law.

    B. Henry's law.

    C. the van't Hoff factor.

    D. the Tyndall effect.

15. The movement of solvent from a region of higher concentration into a region of lower concentration is _____

    A. dialysis.

    B. migration.

    C. desalinization.

    D. osmosis.

16. Solutions that obey Raoult's law are called _____

    A. Raoult solutions.

    B. ideal solutions.

    C. hypertonic.

    D. hypotonic.

17. A mixture that does not separate on standing and that scatters visible light is a(n) _____

    A. suspension.

    B. colloid.

    C. solution.

    D. emulsion.

18. The scattering of visible light by particles in a certain type of mixture is known as _____

    A. micelles.

    B. emulsions.

    C. the photoelectric effect.

    D. the Tyndall effect.

19. A dispersion of solid particles in a liquid or solid is known as a(n) _____

   A. sol.

   B. aerosol.

   C. emulsion.

   D. colloid.

20. Soap contains only hydrophilic parts.

   A. True

   B. False. It only contains hydrophobic parts.

   C. False. It contains both hydrophilic and hydrophobic parts.

Answers: 1. C; 2. B; 3. A; 4. B; 5. D; 6. C; 7. D; 8. C; 9. A; 10. C; 11. A; 12. B; 13. C; 14. A; 15. D; 16. B; 17. B; 18. D; 19. A; 20. C

CHAPTER 14

Chemical Kinetics

Key Words

activated complex	elementary reaction	radioisotope dating
activation energy	elementary reactions	rate constant
activity	enzyme inhibitors	rate-determining step
adsorption	enzymes	rate laws
Arrhenius equation	first-order reaction	rate of decay
average reaction rate	frequency factor	reaction coordinate
bimolecular	half-life	reaction mechanism
carbon-14 dating	heterogeneous catalysis	reaction order
catalysts	homogeneous catalysis	reaction rates
chain-branching steps	hydrogenation	second-order reaction
chain reaction	initiation	substrate
chemical kinetics	instantaneous rate	termination
collision model of	integrated rate law	termolecular
chemical kinetics	intermediate	transition state
desorption	molecularity	unimolecular
differential rate law	propagation	zeroth-order reaction
disintegrations per minute	radical	
disintegrations per second	radioactive decay constant	

By the end of this chapter, you should be able to:

- Describe factors that affect the rate of chemical reactions
- Calculate a reaction rate
- Understand rate laws
- Understand how concentration affects the rate of a reaction
- Describe simple reactions step-wise
- Have a beginning understanding of why chemical reactions occur

Chapter Overview

In previous discussions of chemical behaviors, we've held the assumption that the systems remained constant over time. In this chapter, we shift the discussion to a much more common situation in which the systems change as a function of time. We will talk about reaction rates, which describe the changes in the concentrations of products and reactants as a function of time.

14.1. Factors that Affect Reaction Rates

There are several factors that determine the rate of a reaction, including concentration, temperature, the physical state of the reactants, and the presence of a catalyst. Generally speaking, the reaction rate will increase as the concentration of the reactants increases. The rate also tends to increase as a rise in temperature leads to greater kinetic energy in the particles. If all reactants are in the same phase, this will generally increase the rate, as there will be more area for the reaction to take place. Reactants in different phases will only react at the interface of the phases. Catalysts produce a dramatic increase in the rate of reaction.

14.2. Reaction Rates and Rate Laws

Reaction rates may be calculated for either a single moment of a reaction, giving the instantaneous rate, or for the average over a period of time, giving the average rate. The rate law of a reaction is a mathematical relationship between the reaction rate and the concentration of a species in a solution. Rate laws can be given in two ways: an integrated rate law, which describes the actual concentrations of products or reactants as a function of time; or a differential rate law, which describes the change in product or reactant concentrations as a function of time. The reaction order is the degree to which the rate depends upon the concentration of a reactant and is denoted by the power to which a concentration is raised.

14.3. Methods of Determining Reaction Orders

The reaction order is based upon the extent to which the reaction depends upon the concentration of reactants. In a zeroth-order reaction, the rate is completely independent of the concentrations of the reactants. A plot of concentration versus time for a zeroth-order reaction would be linear. In a first-order reaction, the rate is directly proportional to the concentration of one of the reactants. A plot of a first-order reaction would be linear if the logarithm of the reactant concentration was plotted versus time. In a second-order reaction, the rate is proportional to the concentration of the square of one of the reactants. A plot of a second-order reaction would be linear if the reciprocal of the reactant concentration was plotted versus time.

14.4. Using Graphs to Determine Rate Laws, Rate Constants, and Reaction Orders

A plot of the concentration of any reactant in a zeroth-order reaction versus time will give a straight line with a slope of the negative rate constant, k. A plot of the logarithm of a reactant in a first-order reaction versus time will give a straight line with a slope of the negative rate constant. A plot of the inverse of a reactant in a second-order reaction versus time will give a straight line with a slope of the rate constant.

14.5. Half-Lives and Radioactive Decay Kinetics

The half-life of any reaction is the time required for the reactant concentration to decrease to one-half of its initial value. For a first-order reaction, the half-life is a constant and is related to the rate constant for the reaction. Among first-order reactions is radioactive decay. The rate of decay, or activity, of a sample of radioactive material is the decrease in the number of radioactive nuclei per unit time; this is usually measured in units of disintegrations per second or disintegrations per minute. Radioactive decay has enabled scientists to date artifacts through measuring certain isotopes in the articles, such as carbon-14.

14.6. Reaction Rates—A Microscopic View

A reaction mechanism is the microscopic path by which the reactants are transformed into products. Each step in the mechanism is an elementary reaction and can be described in terms of the number of molecules that collide in that step. A unimolecular step involves only a single reactant. A bimolecular step involves two reactants colliding. And a termolecular step involves three reactants colliding. Species that are formed in one step and consumed in another are called *intermediates*. The slowest step in any reaction mechanism is the rate-determining step. *Chain reactions* are mechanisms that consist of three stages of reactions: initiation, which produces one or more reaction intermediates; propagation, which consumes and reproduces intermediates while products are formed; and termination, which produces stable end products.

14.7. The Collision Model of Chemical Kinetics

The activation energy of a reaction is the minimum energy required for a molecular collision to result in a chemical reaction. At any given temperature, the higher the activation energy is, the slower the reaction. The graph of the potential energy of a system versus the reaction coordinate shows an energy barrier that must be overcome if the reaction is to proceed. The fraction of orientations that can result in a reaction is known as the *steric factor*. The frequency factor, steric factor, and activation energy are related to the rate constant by the Arrhenius equation.

14.8. Catalysis

A catalyst is a species that increases the rate of a chemical reaction without actually being consumed in the reaction. Catalysts increase a reaction's rate by lowering the activation energy of the reaction. Homogeneous catalysts are in the same phase as the reactants, while heterogeneous catalysts are in a different phase and provide a surface to which the reactants can bind in a process called *adsorption*. Enzymes are biological catalysts that are specific for certain products and reactants. The substrate is the reactant in an enzyme-catalyzed reaction. Enzyme inhibitors decrease the rate of these reactions.

Self-Test

1. Which of the following does not influence the rate of a chemical reaction?
   A. Temperature
   B. Concentration
   C. Density
   D. Solvent

2. Increasing the temperature _____ the rate of the reaction.
   A. increases
   B. decreases

3. The proportionality between the reaction rate and the reactant concentration is described by the _____
   A. instantaneous rate.
   B. rate constant.
   C. differential rate law.
   D. reaction order.

4. The change in reactant concentration as a function of time is described by the

   _____

   A. average rate law.
   B. instantaneous rate law.
   C. differential rate law.
   D. integrated rate law.

5. The rate of a _____ reaction is independent of the concentrations of reactants.
   A. zeroth-order
   B. first-order
   C. second-order

6. The rate of a first-order reaction is _____ of the reactants.
   A. independent of the concentration
   B. proportional to the concentration of one
   C. proportional to the concentration of the square of one

7. A plot of reactant concentration versus time for a first-order reaction would be

   _____

   A. linear.
   B. nonlinear.

8. A plot of reactant concentration versus time for a zeroth-order reaction would be _____
   A. linear with a slope of $-k$.
   B. linear with a slope of k.
   C. nonlinear with a slope of $-k$.
   D. nonlinear with a slope of k.

9. A plot of reactant concentration versus time for a second-order reaction would be _____
   A. linear with a slope of $-k$.
   B. linear with a slope of k.
   C. nonlinear with a slope of $-k$.
   D. nonlinear with a slope of k.

10. A plot of the logarithm of a reactant versus time for a first-order reaction would be _____
   A. linear with a slope of $-k$.
   B. linear with a slope of k.
   C. nonlinear with a slope of $-k$.
   D. nonlinear with a slope of k.

11. A first-order reaction has a rate constant of 1.15×10^{-3} min^{-1}. The half-life of the reaction is _____
   A. 205 min.
   B. 331 min.
   C. 462 min.
   D. 525 min.

12. Which of the following is not a radioactive isotope?
   A. Carbon-14
   B. Uranium-238
   C. Potassium-40
   D. Iron-56

13. What is the first-order rate constant of a reaction with a half-life of 630 days?
   A. 1.1×10^{-2} day^{-1}
   B. 1.1×10^{-3} day^{-1}
   C. 1.1×10^{-4} day^{-1}
   D. 1.1 day^{-1}

14. Reaction intermediates are not formed in which step of a chain reaction?
   A. Initiation
   B. Propagation
   C. Termination

15. Which type of elementary reaction is the most rare?

    A. Unimolecular

    B. Bimolecular

    C. Termolecular

16. Which of the following is not related in the Arrhenius equation?

    A. Rate constant

    B. Activation energy

    C. Reaction order

    D. Frequency factor

17. At a given temperature, what effect does a higher activation energy have on a reaction?

    A. Speeds it up

    B. Has no reaction

    C. Slows it down

18. A catalyst increases a reaction rate by _____

    A. raising the temperature.

    B. decreasing the energy barrier.

    C. doubling the reactant concentration.

    D. lowering the solvent viscosity.

19. Adsorption can take place in the presence of _____

    A. any catalyst.

    B. no catalyst.

    C. homogeneous catalysts.

    D. heterogeneous catalysts.

20. Enzymes are primarily composed of _____

    A. heavy metals.

    B. proteins.

    C. hydrogenated molecules.

    D. poisons.

Answers: 1. C; 2. A; 3. B; 4. C; 5. A; 6. B; 7. B; 8. C; 9. D; 10. A; 11. C; 12. D; 13. B; 14. C; 15. C; 16. C; 17. C; 18. B; 19. D; 20. B

CHAPTER 15

Chemical Equilibrium

Key Words

chemical equilibrium	heterogeneous equilibrium	physical equilibrium
equilibrium constant	homogeneous equilibrium	reaction quotient
equilibrium constant expression	kinetic control	thermodynamic control
equilibrium equation	law of mass action	
	Le Chatelier's principle	

By the end of this chapter, you should be able to:

- Define *chemical equilibrium*
- Relate the equilibrium constant to the rate constant for both directions of a reaction
- Write an equilibrium constant expression for any reaction
- Solve chemical equilibrium problems
- Predict the direction of a reaction
- Predict the effects of outside influences on a system's equilibrium
- Understand how to control the products of a reaction

Chapter Overview

We've spent quite a bit of time talking about chemical reactions. In the last chapter, we discussed kinetics, which dealt with the rates of reactions. In this chapter, we will discuss chemical equilibrium, which is a description of the extent to which a chemical reaction occurs. Virtually all reactions are reversible, in that an opposing reaction can take place to re-form the reactants from the products. Equilibrium describes the point at which the rates of the forward and reverse reactions are equal.

15.1. The Concept of Chemical Equilibrium

Chemical equilibrium is defined as a dynamic process consisting of forward and reverse reactions that proceed at equal rates. At the point of equilibrium, the composition of the system is static with time. The composition of an equilibrium mixture is the same, regardless of the direction from which it is approached.

15.2. The Equilibrium Constant

The equilibrium constant is the ratio of rate constants for the forward and reverse reactions of a system at equilibrium. The equilibrium constant expression is the ratio of the equilibrium concentrations of the products and reactants, raised to their respective coefficients. When a reaction is written in the reverse direction, the equilibrium constant and equilibrium constant expression are inverted. A system at equilibrium that contains both products and reactants that are all in a single phase is said to be in a state of homogeneous equilibrium. A system with reactants, products, or both in more than one phase is said to be in a state of heterogeneous equilibrium. When a reaction can be written as a sum of two or more reactions, its equilibrium constant is equal to the product of the equilibrium constants for the individual reactions.

15.3. Solving Equilibrium Problems

The equilibrium constant may be calculated from the equilibrium concentrations by substituting molar concentrations or partial pressures into the expression for the reaction. Equilibrium concentrations may also be calculated from equilibrium constants by using a tabular format:

$$X \rightleftharpoons Y$$

	[X]	[Y]
Initial		
Change		
Final		

1. Calculate the initial concentrations for all species possible from the data given
2. Use the coefficients in the balanced equation to find the changes in all other species in the reaction
3. Obtain the final concentrations by summing the initial and change rows
4. Calculate the equilibrium constant: $K = \dfrac{[Y]}{[X]}$

15.4. Nonequilibrium Conditions

The reaction quotient is derived from concentrations obtained at any time and is equal to the equilibrium constant if those concentrations are obtained when the system is at equilibrium. The direction in which a reaction will proceed can be predicted by plotting a few equilibrium concentrations at a given temperature and pressure. Points that lie outside the line established in the plot represent nonequilibrium states, and the system will readjust itself to establish equilibrium.

15.5. Factors that Affect Equilibrium

Le Chatelier's principle states that if a stress is applied to a system at equilibrium, the composition of the system will adjust to relieve that stress. There are three things that can alter a reaction in this way: (1) a change in temperature, (2) a change in pressure, or (3) a change in the concentrations of either reactants or products. It is possible to manipulate the conditions of a reaction in order to drive an unfavorable reaction to completion. Lowering the temperature of an exothermic reaction will shift it toward the product's side, while lowering the temperature of an endothermic reaction will shift it toward the reactant's side. The reverse also holds true. Raising the pressure of a reaction involving gaseous substances will favor the side with the fewest gaseous molecules.

15.6. Controlling the Products of Reactions

It is possible to control both the rate and the outcome of a chemical reaction. Reaction conditions may be adjusted to control which product or products are present at equilibrium. This is known as *thermodynamic control*. The reaction rate may be adjusted to obtain a single product. This is known as *kinetic control*.

Self-test

1. Which of the following is in a state of equilibrium?

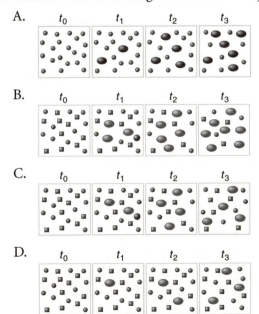

2. In a state of chemical equilibrium, the rate of the reaction forming the products is _____ the rate of the reaction forming the reactants.

 A. less than

 B. equal to

 C. greater than

3. Write the equilibrium equation for the following reaction:

$$2 \; ZnS + 3 \; O_2 \rightarrow 2 \; ZnO + 2 \; SO_2$$

 A. $K = \dfrac{[ZnS]^2[O]^2}{[ZnO]^2[SO_2]^2}$

 B. $K = \dfrac{[ZnO]^2[SO_2]^2}{[ZnS]^2[O]^2}$

 C. $K = \dfrac{[ZnS]^2[O_2]^3}{[ZnO]^2[SO_2]^2}$

 D. $K = \dfrac{[ZnO]^2[SO_2]^2}{[ZnS]^2[O_2]^3}$

4. Write the equilibrium equation for the reverse of the following reaction:

$$2 \; NOCl \rightarrow 2 \; NO + Cl_2$$

 A. $\dfrac{[NOCl]^2}{[NO]^2[Cl]^2}$

 B. $\dfrac{[NO]^2[Cl]^2}{[NOCl]^2}$

 C. $\dfrac{[NOCl]^2}{[NO]^2[Cl_2]}$

 D. $\dfrac{[NO]^2[Cl_2]}{[NOCl]^2}$

5. A certain reaction has an equilibrium constant of 0.5. Calculate the constant for the reverse reaction.

 A. 0.25

 B. 0.5

 C. 1

 D. 2

6. The following equation is an example of what kind of equilibrium?

$$CO_2(g) + C(s) \rightarrow 2 \; CO_2(g)$$

 A. Physical equilibrium

 B. Nonequilibrium

 C. Homogeneous equilibrium

 D. Heterogeneous equilibrium

7. A certain reaction has an equilibrium constant of 3.2×10^{-4} under experimental conditions. The reaction quotient of the same reaction is determined to be 3.05×10^{-4}. In which direction will the reaction proceed?

   A. To the right

   B. To the left

   C. Neither; the reaction is at equilibrium.

8. Write the reaction quotient expression for the following reaction:

   $$2 \, ZnS + 3 \, O_2 \rightarrow 2 \, ZnO + 2 \, SO_2$$

   A. $K = \dfrac{[ZnS]^2[O]^2}{[ZnO]^2[SO_2]^2}$

   B. $K = \dfrac{[ZnO]^2[SO_2]^2}{[ZnS]^2[O]^2}$

   C. $K = \dfrac{[ZnS]^2[O_2]^3}{[ZnO]^2[SO_2]^2}$

   D. $K = \dfrac{[ZnO]^2[SO_2]^2}{[ZnS]^2[O_2]^3}$

9. A certain reaction has an equilibrium constant of 1.5×10^{-3}. What is the reaction quotient of the same reaction at equilibrium?

   A. 0.15

   B. 0.015

   C. 0.0015

   D. 0.00015

10. Consider the following reaction:

    $$N_2 + 3 \, H_2 \rightarrow 2 \, NH_3$$

    Given that ammonia has a partial pressure of 1.74, nitrogen has a partial pressure of 6.59, and hydrogen has a partial pressure of 21.58, calculate the reaction quotient for the reaction.

    A. 3.83×10^{-5}

    B. 6.9×10^{-6}

    C. 4.57×10^{-5}

    D. 5.12×10^{-6}

11. The tendency of a system to adjust itself to compensate for stresses upon the system is known as _____

    A. a nonequilibrium condition.

    B. Le Chatelier's principle.

    C. kinetic control.

    D. thermodynamic control.

12. Consider the following reaction:

$$CuO(s) + CO(g) \rightarrow Cu(s) + CO_2(g)$$

What effect would increasing the amount of CO have on the products?
A. It would raise the amount of both products.
B. It would raise the amount of Cu.
C. It would raise the amount of CO_2.
D. It wouldn't have any effect on the products.

13. What effect does raising temperature have on an endothermic reaction?
A. Shifts the reaction toward reactants
B. Shifts the reaction toward products
C. Has no effect

14. Consider the following reaction:

$$2\ H_2(g) + O_2(g) \rightarrow 2\ H_2O(g)$$

What effect would doubling the pressure have on the reaction?
A. Shifts the reaction toward the reactants
B. Shifts the reaction toward the products
C. Has no effect

15. Adjusting reaction rates to control the outcome of a reaction is known as _____
A. kinetic control.
B. thermodynamic control.

Answers: 1. C; 2. B; 3. D; 4. C; 5. D; 6. D; 7. A; 8. D; 9. C; 10. C; 11. B; 12. A; 13. B; 14. B; 15. A

CHAPTER 16

Aqueous Acid–Base Equilibria

Key Words

acid–base indicator
acid ionization
  constant (K_a)
amphiprotic
analytical concentration
Arrhenius acid/base
base ionization
  constant (K_b)
Brønsted–Lowry
  acid/base

buffers
common ion effect
conjugate acid–base pair
equivalence point
Henderson–Hasselbalch
  equation
hydrolysis reaction
hydronium ion
ion-product constant of
  liquid water (K_w)

leveling effect
Lewis acid/base
midpoint
neutralization reaction
pH scale
pOH scale
polyprotic acid/base
titrant
titration curve

By the end of this chapter, you should be able to:

- Recognize the amphiprotic quality of water
- Understand the relationship between pH, pOH, and pK_w
- Recognize conjugate acid–base pairs
- Understand K_a, pK_a, K_b, and pK_b and how they relate to the strength of an acid or base
- Understand acid–base equilibria
- Calculate the percent ionization of a solution
- Calculate pH of a solution
- Construct a titration curve and identify the equivalence point
- Understand how buffers work
- Use the Henderson–Hasselbalch equation

Chapter Overview

In the last chapter, we developed the concept of chemical equilibrium. In this chapter, we continue our quantitative look at chemical reactions by developing concepts and equations relating to acid–base reactions.

16.1. The Autoionization of Water

Water is an amphiprotic substance, meaning that it can act as either an acid or a base. It acts as an acid when it donates protons to a base to form hydroxide ions, OH^-. It acts as a base when it accepts protons to form hydronium ions, H_3O^+. Water is capable of autoionizing, in which both OH^- and H_3O^+ are produced. The chemical equilibrium for this reaction is known as the *ion-product constant of liquid water* and is given the notation K_w. K_w at 25°C is equal to 1.0×10^{-14}.

16.2. A Qualitative Description of Acid–Base Equilibria

The relative strength of an acid or base can be found from the magnitude of the equilibrium constant for an ionization reaction of that species. This dissociation constant is known as the *acid ionization constant*, K_a, for a weak acid. For a weak base, the dissociation constant is known as the *base ionization constant*, K_b. A conjugate acid–base pair is two species that differ from each other by only a proton. For any conjugate acid–base pair, $K_a \times K_b = K_w$. The pK values for acids or bases are found by taking the negative log of the K_a or K_b values, respectively. The smaller the pK value, the stronger the acid or base is. At 25°C, pK_a + pK_b = pK_w = 14.00. Acid–base reactions will always proceed in the direction that yields the weakest acid–base pair, and hence the highest pK values.

Polyprotic acids and bases, like their name implies, have more than one proton present in the species. The fully protonated form is the strongest acid, while the fully deprotonated form is the strongest base. An aqueous salt solution can be acidic if it contains the conjugate base of a weak acid as the anion. It can be basic if it contains the conjugate acid of a weak base as the cation. Or, it can be neutral if it contains both. Salts that contain small, highly charged metal ions will produce acidic solutions in water. A hydrolysis reaction occurs when a salt reacts with water to produce an acidic or basic solution.

16.3. Molecular Structure and Acid–Base Strength

The strength of an acid or base depends on its structure. The weaker the bond between the proton and the rest of the molecule, the more likely it will be to dissociate and form an H^+ ion. The stronger the attraction between the lone pair and a conjugate base, the more favorable dissociation of the H^+ will be and the stronger the resulting conjugate acid will be. The inductive effect occurs when an atom or group of atoms elsewhere on the molecule weakens an O—H bond, which allows H^+ to be formed more easily.

16.4. Quantitative Aspects of Acid–Base Equilibria

If we know the concentration of one or more of the species present in a solution of an acid or a base, then we can calculate K_a and K_b. The values of K_a, K_b, pK_a, and pK_b are useful in describing the composition and strength of acids and bases. The concentrations of all the species present in an acid or base can be determined mathematically, as can the pH of the solution and the percent ionization of the acid or base. The equilibrium constant for the reaction of a weak acid with a weak base can be calculated from K_a (or pK_a), K_b (or pK_b), and K_w.

16.5. Acid–Base Titrations

A titration curve is constructed by plotting the volume of titrant added versus the pH of a solution during an acid–base titration. The overall shape of the resulting curve can provide a good deal of information about what is occurring in the solution during the titration. The shapes of the curves for weak acids and bases differ widely and depend on the compound being titrated. The equivalence point in a titration curve is the point at which exactly enough acid or base has been added to neutralize the other component. The equivalence point for a strong acid or base will occur at a pH of 7.00. The equivalence point for a weak acid will occur at a pH greater than 7.00, while the point for a weak base will occur at a pH less than 7.00. The pH of the titration of a weak acid or base will tend to rise more slowly before the equivalence point, while it will rise more sharply for strong acids and bases. The midpoint of a titration curve is the point that lies halfway to the equivalence point, and the pH at that point is equal to the pK_a of a weak acid or the pK_b of a weak base. Titrations can be followed visually through the use of an acid–base indicator, which is usually a weak acid or a weak base that changes color in response to the protonation or deprotonation of the indicator.

16.6. Buffers

A buffer is a solution that resists a change in pH upon the addition of an acid or a base. A buffer contains either a weak acid and its conjugate base, or a weak base and its conjugate acid. By adding a strong electrolyte that contains one ion in common with a reaction system that is in equilibrium, the equilibrium can be shifted in the direction that reduces the concentration of the common ion. This is referred to as the *common ion effect*. A buffer may be characterized by their buffer capacity and their pH range. The pH range of a buffer depends upon the pK_a or pK_b of the acid or base used to prepare the buffer. The buffer capacity depends upon the concentrations of the species in the solution. The pH of a buffer can be calculated by using the Henderson–Hasselbalch equation.

Self-Test

1. Having the ability to act as both an acid and a base is known as being _____
   A. autoionized.
   B. amphiprotic.
   C. hydrolyzed.
   D. leveled.

2. The ion product constant of water is equal to _____
   A. 1×10^{-7}.
   B. 7.00.
   C. 1×10^{-14}.
   D. 14.00.

3. The conjugate acid of ammonia is _____
   A. NH_3.
   B. NH_2^-.
   C. NH_4^+.
   D. HNH_2.

4. The conjugate base of sulfuric acid is _____
   A. H_2SO_4.
   B. HSO_4^-.
   C. H_2SO_3.
   D. SO_4^{-2}.

5. Acetic acid has a pK_a of 4.76. What is the K_a value for this acid?
   A. 2.0×10^{-5}
   B. 4.76×10^{-5}
   C. 3.2×10^{-5}
   D. 1.7×10^{-5}

6. Sulfuric acid has a K_a value of 1.0×10^{-2}. What is the pK_b value for this acid?
   A. 14
   B. 16
   C. 12
   D. 10

7. Rank the following acids in terms of increasing strength: HF, HCl, HBr, and HI.
   A. HI < HBr < HCl < HF
   B. HI < HCl < HBr < HF
   C. HF < HCl < HBr < HI
   D. HF < HBr < HCl < HI

8. What is the K_a expression for the dissociation of sulfuric acid?
   A. $K_a = \dfrac{[H^+][SO_4]}{[H_2SO_4]}$
   B. $K_a = \dfrac{[H_2SO_4]}{[H^+][SO_4^-]}$

9. Calculate the concentration of H^+ if 0.15 M HNO$_3$ ($K_a = 2.3 \times 10^1$) dissociates.

A. 0.58 M

B. 1.85 M

C. 3.92 M

D. 2.14 M

10. Calculate the percent ionization if 0.75 M acetic acid ($K_a = 1.7 \times 10^{-5}$) dissociates.

A. 0.6%

B. 0.5%

C. 0.4%

D. 0.3%

11. What is the pH of an acid with $[H^+] = 3.45 \times 10^{-3}$?

A. 4.68

B. 3.13

C. 2.46

D. 1.99

12. What volume of 0.15 M NaOH is needed to titrate 25 mL of 0.12 M HCl?

A. 12 mL

B. 18 mL

C. 25 mL

D. 20 mL

13. The equivalence point of a titration of a weak acid would be _____

A. less than 7.00.

B. 7.00.

C. more than 7.00.

14. What is the pH at the midpoint of a titration of acetic acid ($K_a = 1.7 \times 10^{-5}$)?

A. 3.34

B. 4.76

C. 5.16

D. 6.02

15. Calculate the pH of a buffer solution that is 0.040 M Na_2HPO_4 and 0.080 KH_2PO_4, given that the K_a for H_2PO_4 = 7.21.

    A. 6.91

    B. 6.72

    C. 7.13

    D. 7.47

Answers: 1. B; 2. C; 3. C; 4. B; 5. D; 6. B; 7. C; 8. A; 9. B; 10. B; 11. C; 12. D; 13. C; 14. B; 15. A

CHAPTER 17

Solubility and Complexation Equilibria

Key Words

acidic oxides

amphoteric oxides

basic oxides

complexing agents

complex ion

eutrophication

formation constant

incomplete dissociation

ion pair

ion product

ligands

qualitative analysis

scale

solubility product

stalactite

stalagmite

By the end of this chapter, you should be able to:

- Calculate the solubility of an ionic compound from its solubility product
- Understand the factors that determine the solubility of ionic compounds
- Describe the formation of a complex ion
- Understand the dependence of solubility on pH
- Use a qualitative analysis experiment to separate metal ions

Chapter Overview

In the previous chapter, we looked at the equilibria of acid–base reactions. In this chapter, we will discuss solubility equilibria, in which certain reactions may lead to the formation of a solid precipitate.

17.1. Determining the Solubility of Ionic Compounds

The solubility product, K_{sp}, is the equilibrium constant for a dissociation reaction and measures the solubility of a compound. The solubility of a compound is usually expressed in terms of the mass of solute per 100 mL of solvent, while the solubility product is expressed in terms of the molar concentrations of the component ions. The ion product, Q, describes concentrations that are not necessarily the equilibrium concentrations of an ionic compound. The comparison of Q to K_{sp} will help you determine whether or not a precipitate will form if solutions of two soluble salts are mixed. The addition of a common cation or anion to a solution of sparingly soluble salts will shift the equilibrium, almost always decreasing the solubility of the salt.

17.2. Factors Affecting Solubility

There are four things that can affect the solubility of a compound. One is ion pair formation, in which an anion and a cation interact with each other and are not separated in solution by a solvent. The second is the incomplete dissociation of molecular solutes. The third is the formation of complex ions, and the fourth is a change in solution pH. An ion pair is the result of electrostatic attractive forces between the cation and the anion. Incomplete dissociation is the result of intramolecular forces not involving the cation and anion.

17.3. Complex Ion Formation

A complex ion is formed when molecules or ions (ligands) that contain one or more lone pairs of electrons surround a central metal ion. Complex ions will most likely be formed from small, highly charged metal ions than from those that are larger or not as highly charged. The equilibrium constant for the formation of the complex ion is given by the formation constant, K_f. The formation of a complex ion has a tendency to increase the solubility of a compound.

17.4. Solubility and pH

The anion in many sparingly soluble salts is the conjugate base of a weak acid. As the pH is lowered, that anion will become protonated, which can dramatically increase the solubility of the salt. Oxides can be either acidic or basic. Acidic oxides dissolve in strong bases or react with water to give acidic solutions. Basic oxides dissolve in strong acids, or react with water to give basic solutions. Most acidic oxides are either nonmetal oxides or else metal oxides with high oxidation states, while basic oxides tend to be metallic. Some oxides are capable of dissolving in both acidic and basic solutions, making them amphoteric. Most amphoteric oxides come from the semimetallic elements.

17.5. Qualitative Analysis Using Selective Precipitation

The identity of metal ions present in a mixture can be determined through a procedure known as *qualitative analysis*. In a qualitative analysis, metal ions are selectively precipitated a few at a time. Consecutive precipitation steps become progressively less selective until almost all of the metal ions are precipitated. If metal ions precipitate together, then other methods of separation must be used.

Self-Test

1. Under what condition will a precipitate form?
   A. $Q < K_{sp}$
   B. $Q = K_{sp}$
   C. $Q > K_{sp}$

2. What is the K_{sp} expression for the following dissociation reaction?
   A. $K_{sp} = [Pb^{2+}] [Cl^-]$
   B. $K_{sp} = [Pb^{2+}]^2 [Cl^-]$
   C. $K_{sp} = [Pb^{2+}] [Cl^-]^2$
   D. $K_{sp} = [Pb^{2+}]^2 [Cl^-]^2$

3. The solubility of a salt is generally _____ by the presence of a common ion.
   A. increased
   B. decreased
   C. unaffected

4. Which of the following would not change the solubility of a compound?
   A. Ion pair formation
   B. Use of powdered solids instead of granules
   C. Formation of complex ions
   D. Changes in pH

5. The actual molar solubility of $Mg(OH)_2$ is expected to be _____ than the value calculated from its K_{sp}.
   A. higher
   B. the same
   C. lower

6. The actual molar solubility of a salt in acetic acid solution is expected to be _____ than the value calculated from its K_{sp}.
   A. higher
   B. the same
   C. lower

7. What is the K_f expression for the complex ion formed from Co(II) and (C_5H_5N)?

A. $K_f = \dfrac{[Co(C_5H_5N)]}{[Co(II)]\,[C_5H_5N]}$

B. $K_f = [Co(II)]\,[C_5H_5N]$

C. $K_f = \dfrac{[Co(II)]\,[C_5H_5N]}{[Co(C_5H_5N)]}$

D. $K_f = [Co(C_5H_5N)]$

8. The actual molar solubility of a salt that forms a complex ion is expected to be _____

A. higher than the value calculated from its K_{sp}.
B. the same as the value calculated from its K_{sp}.
C. lower than the value calculated from its K_{sp}.
D. dependent upon the K_f of the complex ion.

9. The use of phosphates in detergents can harm the environment by causing an overgrowth of algae known as _____

A. complexation.
B. eutrophication.
C. blooming.
D. spirulina.

10. Decreasing the pH of a solution will have what effect on the solubility of salts in the solution?

A. Increasing
B. No effect
C. Decreasing
D. Depends on the composition of the salt

11. Rank the following salts in order of increasing solubility in acidic solution:

A. $AgBr < AgCl < AgCO_3 < AgPO_4$
B. $AgCl < AgBr < AgPO_4 < AgCO_3$
C. $AgBr < AgCO_3 < AgCl < AgPO_4$
D. $AgBr < AgCl < AgPO_4 < AgCO_3$

12. CrO would be classified as a(n) _____ oxide.

A. acidic
B. basic
C. amphoteric

13. A metal chloride is found to precipitate in water. Which of the following cations might be present?

    A. Cu^{2+}

    B. Fe^{2+}

    C. Pb^{2+}

    D. Cd^{2+}

14. How might one test for alkali metals in a salt solution?

    A. Addition of NaOH

    B. Addition of HCl

    C. Formation of a complex ion

    D. Flame test

15. The process of determining the identity of metal ions present in a mixture is known as _____

    A. complex ion formation.

    B. salting out.

    C. qualitative analysis.

    D. quantitative analysis.

Answers: 1. C; 2. C; 3. B; 4. B; 5. A; 6. A; 7. A; 8. D; 9. B; 10. D; 11. B; 12. B; 13. C; 14. D; 15. C

CHAPTER 18

Chemical Thermodynamics

Key Words

enthalpy	irreversible process	standard free energy of
entropy	photosynthesis	formation
fermentation	respiration	standard molar entropy
first law of	reversible process	state function
thermodynamics	second law of	thermodynamics
Gibbs free energy	thermodynamics	third law of
internal energy	standard free-energy change	thermodynamics

By the end of this chapter, you should be able to:

- Understand how work, heat, and energy are related to each other
- Define the concept of PV work
- Calculate changes in internal energy
- Understand how entropy is related to internal energy
- Use a thermodynamic cycle to calculate changes in entropy
- Understand the relationship between Gibbs free energy and work
- Understand the information about a system that thermodynamics can provide
- Understand the role thermodynamics plays in biochemical systems

Chapter Overview

Chemical reactions obey two fundamental laws: the law of conservation of mass and the law of conservation of energy. In an earlier chapter, we explored the thermochemistry of a reaction. In this chapter, we develop the idea of thermodynamics: the interrelationships between heat, work, and the energy of a system.

18.1. Thermodynamics and Work

Thermodynamics is defined as the study of the interrelationships among heat, work, and the energy content of a system at equilibrium. The internal energy of a system is the sum of the potential and kinetic energy of all the components of the system. When the pressure or volume of a gas is changed, any mechanical work done as a result of this change is called *PV work*. Any work done by the system on the surroundings is considered to be negative, while any work done on the system by the surroundings is considered to be positive.

18.2. The First Law of Thermodynamics

The first law of thermodynamics says that the energy in the universe is constant. The change in a system's internal energy is the sum of the heat transferred and the work done. *Enthalpy (H)* is defined as the relationship between the heat flow (q) and the internal energy (E) of a system at constant pressure. Heat flow is the change in the internal energy of the system plus any *PV* work done.

18.3. The Second Law of Thermodynamics

The second law of thermodynamics says that for a reversible process, the entropy of the universe is constant, whereas in an irreversible process, the entropy of the universe increases. Entropy (s) is the measure of the disorder of a system. A process is considered reversible if all the intermediate states of the process are equilibrium states, making it possible for the reaction to change direction at any time. An irreversible process, however, occurs in only one direction. The change in entropy is the heat transferred into or out of a system, divided by the temperature.

18.4. Entropy Changes and the Third Law of Thermodynamics

The third law of thermodynamics says that the entropy of any perfectly ordered, crystalline substance at absolute zero is zero. This means that any substance above absolute zero has a positive entropy. Changes in entropy can be calculated by measuring the heat capacity of the substance and using the enthalpies of fusion or vaporization. The standard molar entropy is defined as the entropy of 1 mol of a substance at a standard temperature of 298 K. The standard entropy change for a reaction can be calculated from tabulated values of standard molar entropies by subtracting the reactants from the products.

18.5. Free Energy

The enthalpy, entropy, and temperature of a system can be combined into a function known as *Gibbs free energy* (G), which can be used to predict whether a reaction will occur spontaneously. The change in free energy (ΔG) is the difference between the heat released during a reversible or irreversible process and the heat released from the same process occurring as reversible. A system at equilibrium will have a change in free energy equal to zero. For a spontaneous reaction, the change in free energy will be negative. For a reaction that is not spontaneous as written, the change in free energy will be positive. If the temperature and pressure are held constant, the change in free energy will be equal to the maximum amount of work that a system can perform on the surroundings while undergoing a spontaneous change. The *standard free-energy change* ($\Delta G°$) is defined as the change in free energy when one substance or more, in their standard states, is converted to one or more other substances, also in their standard states. The *standard free energy of formation* ($\Delta G_f°$) is the change in free energy that occurs when 1 mol of a substance in its standard state is formed from the elements in their standard states.

18.6. Spontaneity and Equilibrium

The change in free energy can be expressed in terms of volume, pressure, entropy, and temperature if the process is reversible and no external work is done on the system. If ideal gas behavior is assumed, then the ideal gas law can be used to express the change in free energy in terms of the partial pressures of the products and reactants. If $\Delta G°$ is less than zero, then the products are favored over the reactants. If $\Delta G°$ is greater than zero, then reactants are favored over products. And if $\Delta G°$ is equal to zero, then the reaction is at equilibrium. The equilibrium constant at one temperature and the standard enthalpy of formation can be used to estimate the equilibrium constant for the reaction at any other temperature.

18.7. Comparing Thermodynamics and Kinetics

Kinetics is used to describe the rate at which a particular process will occur and the pathway along which it will occur. Thermodynamics is used to describe a system's overall properties, behavior, and equilibrium composition. Thermodynamics will tell us what can possibly occur during a process, while kinetics tells us what actually does occur at an atomic and molecular level. A reaction that is not thermodynamically spontaneous under standard conditions can often be made to occur spontaneously by varying the experimental conditions, using a different reaction, or by supplying external energy.

18.8. Thermodynamics and Life

A living cell is a system that is not in equilibrium with its surroundings and requires a constant input of energy to maintain that nonequilibrium. Organisms may be either aerobic or anaerobic, with aerobic organisms unable to survive without oxygen and anaerobic organisms unable to survive in the presence of oxygen. Green plants and algae gain energy from their surroundings by photosynthesis. Other species gain energy from chemical compounds through processes such as respiration or fermentation.

Self-Test

1. The study of thermodynamics does not include _____
   A. heat.
   B. work.
   C. ionization.
   D. energy.

2. Given a frictionless piston with external pressure greater than internal pressure, the value of work done would be _____
   A. positive.
   B. zero.
   C. negative.

3. In a closed container, enthalpy is the relationship between internal energy of a system and _____
   A. PV work.
   B. heat flow.
   C. entropy.
   D. state functions.

4. The change in internal energy of a system is a state function.
   A. True
   B. False
   C. Depends

5. The work done in a reversible process is always _____ the work done in a corresponding irreversible process.
   A. less than
   B. less than or equal to
   C. greater than
   D. greater than or equal to

6. The measure of a system's disorder is its _____
   A. enthalpy.
   B. entropy.
   C. fusion.
   D. heat flow.

7. Ice has a heat of fusion of 6.01 kJ/mol. Calculate the entropy of fusion of 1 mol of ice.

   A. 6.01 kJ/(mol K)

   B. 11 kJ/(mol K)

   C. 22 J/(mol K)

   D. 12.02 kJ/(mol K)

8. According to the third law of thermodynamics, the entropy of any perfectly ordered system at absolute zero is _____

   A. positive.

   B. zero.

   C. negative.

9. Calculate $\Delta S°$ for the formation of water from hydrogen and oxygen given that $S°$ (O_2) = 205.2 J/(mol K); $S°$ (H_2) = 130.7 J/(mol K); and $S°$ (H_2O) = 70.0 J/(mol K).

   A. -326.6 J/(mol K)

   B. -396.6 J/(mol K)

   C. -195.9 J/(mol K)

   D. -275.2 J/(mol K)

10. Which of the following would you expect to have the highest entropy?

    A. NaCl

    B. C_8H_{18}

    C. H_2O

    D. CH_3OH

11. For a spontaneous process, ΔG would be _____

    A. positive.

    B. zero.

    C. negative.

12. Calculate the ΔG of 1 mol of a reaction at STP if ΔH is equal to 250.0 kJ/mol and ΔS is equal to 52.2 kJ/(mol K).

    A. -14903.8 kJ/mol

    B. -14307.5 kJ/mol

    C. -15305.6 kJ/mol

    D. -14000.6 kJ/mol

13. The change in free energy that occurs when 1 mol of a substance in its standard state is formed from the elements in their standard states is called the _____

    A. Gibbs free energy.

    B. standard free-energy change.

    C. standard free energy of formation.

    D. free energy of formation.

14. Products are favored over reactants in a process when _____

    A. $\Delta G° < 0$ and $K > 1$.

    B. $\Delta G° = 0$ and $K = 1$.

    C. $\Delta G° > 0$ and $K < 1$.

    D. $\Delta G° < 0$ and $K < 1$.

15. Reactants are favored over products in a process when _____

    A. $\Delta G° < 0$ and $K > 1$.

    B. $\Delta G° = 0$ and $K = 1$.

    C. $\Delta G° > 0$ and $K < 1$.

    D. $\Delta G° > 0$ and $K > 1$.

16. Thermodynamics describes the pathway by which a reaction will occur.

    A. True

    B. False

    C. Depends

17. Which of the following is not a way to make a thermodynamically nonspontaneous reaction occur spontaneously?

    A. Supplying external energy

    B. Varying reaction conditions

    C. Making the free energy positive

    D. Using a different reaction to obtain the product

18. Green plants generate energy through a process known as _____

    A. digestion.

    B. fermentation.

    C. respiration.

    D. photosynthesis.

19. Organisms that use chemical species such as sulfates as oxidants generate energy through a process known as _____

    A. photosynthesis.

    B. aerobic respiration.

    C. anaerobic respiration.

    D. fermentation.

20. The primary energy source of the cell is _____
    A. NADH.
    B. ATP.
    C. FADH.
    D. NAD$^+$.

Answers: 1. C; 2. A; 3. B; 4. A; 5. D; 6. B; 7. C; 8. B; 9. A; 10. B; 11. C; 12. D; 13. C; 14. A; 15. C; 16. B; 17. C; 18. D; 19. C; 20. B

CHAPTER 19

Electrochemistry

Key Words

- alkaline battery
- amperes
- anode
- battery
- cathode
- corrosion
- coulombs
- disposable (primary) battery
- electrochemical cell
- electrochemistry
- electrodes
- electrolysis
- electroplating
- faraday

- fuel cell
- galvanic cell
- glass electrode
- half-reactions
- indicator electrode
- ion-selective electrodes
- lead-acid battery
- lithium-iodine battery
- Nernst equation
- nickel-cadmium battery
- overvoltage
- oxidant
- potential
- rechargeable (secondary) battery

- reductant
- reference electrode
- salt bridge
- saturated calomel electrode (SCE)
- silver-silver chloride electrode
- standard cell potential
- standard electrode potential
- standard hydrogen electrode (SHE)
- standard reduction potentials
- voltaic cell

By the end of the chapter, you should be able to:

- Distinguish between galvanic and electrolytic cells
- Use redox potentials to predict spontaneous reactions
- Use half-reactions to balance redox reactions
- Understand the relationship between cell potential and the equilibrium constant
- Use cell potentials to measure solution concentrations
- Understand how commercial galvanic cells operate
- Describe the process of corrosion
- Quantitatively describe electrolysis

Chapter Overview

Earlier in the book, we discussed several different types of reactions, including redox reactions. In a redox reaction, one species is reduced while another is oxidized. The transfer of electrons that occurs during a redox reaction is often accompanied by a generation of electrical energy. In this chapter, we discuss these electrochemical reactions in detail and explore some of their applications.

19.1. Describing Electrochemical Cells

Electrochemistry is the study of the relationship between chemical reactions and electricity. Electrochemical processes occur through oxidation–reduction reactions, which can be broken down into two half-reactions—one for the oxidation reaction and another for the reduction reaction. The overall reaction is said to be balanced when the number of electrons lost by the reductant is equal to the number of electrons gained by the oxidant. The flow of electrons from the reductant to the oxidant produces the electrical current. An electrochemical cell is capable of generating electricity from a spontaneous reaction or consuming electricity in order to drive a nonspontaneous reaction. A galvanic cell will only generate electricity from a spontaneous reaction, whereas an electrolytic cell will only consume electricity to drive a reaction to completion. All cells contain two electrodes, with the oxidation occurring at the anode and the reduction occurring at the cathode. A salt bridge connects separated solutions, allowing ions to migrate to either solution in order to ensure electrical neutrality. A voltmeter can be used to measure the flow of electrical current between the two half-reactions. The potential is measured in volts and is the energy needed to move a charged particle in an electric field.

19.2. Standard Potentials

The flow of electrons in a cell depends on several things: the concentrations of the reacting substances, the identities of these substances, and the difference in potential energy of their valence electrons. The standard cell potential, E°_{cell}, is the potential of a cell under standard conditions and a fixed temperature of 25°C. It is the difference between the potentials of two electrodes that is measured; such values of standard electrode potentials are referred to as *standard reduction potentials*. The potential of the standard hydrogen electrode (SHE) is defined as 0 V, and the potential of any half-reaction measured against the SHE is its *standard electrode potential*.

To balance a redox reaction, use the following steps:

1. Write the reduction half-reaction and the oxidation half-reaction.
2. Balance the atoms by balancing elements other than O and H.
3. Balance O atoms by adding H_2O.
4. Balance H atoms by adding H^+.
5. Balance the charges in each half-reaction by adding electrons.
6. Multiply the reductive and oxidative half-reactions by integers to obtain the same number of electrons in both half-reactions.
7. Add the two half-reactions and cancel substances that appear on both sides of the equation.
8. Check that all atoms and charges are balanced.

The potential of an indicator electrode responds to the concentration of the substances being measured, while the potential of the reference electrode is held constant. The potential of the sample versus the potential of the reference electrode will determine whether oxidation or reduction takes place. There are several types of reference electrodes: the standard hydrogen electrode (SHE), the silver-silver chloride electrode, the saturated calomel electrode (SCE), the glass electrode, and ion-selective electrodes.

19.3. Comparing Strengths of Oxidants and Reductants

Standard reduction potentials can be used to compare the oxidative and reductive strengths of several substances. This can lead to departures from expected patterns. These anomalies can occur because the reduction potentials are measured in aqueous solution, which allows for strong intermolecular electrostatic interactions. Such interactions would not be seen in the gas phase.

19.4. Electrochemical Cells and Thermodynamics

Electrochemical current is measured in the metric unit of coulombs (C), which is defined as the number of electrons that pass a given point in 1 s. A current of one ampere (A) is the flow of 1 C/s past a given point. The faraday (F) corresponds to the charge on 1 mol of electrons. Spontaneous redox reactions have a negative ΔG and a positive E°_{cell}. The Nernst equation is perhaps the most often used equation in electrochemistry and can be used to determine the spontaneous direction of any redox reaction under any reaction conditions from values of the relevant standard reduction potentials. Specialized cells known as *concentration cells* have different concentrations of reactants in the anode and cathode compartments and can be used to look at reaction equilibrium. Galvanic cells can be used to measure both the concentration of a species and the solubility product of a sparingly soluble salt.

19.5. Commercial Galvanic Cells

A battery is a self-contained unit that produces electricity. A fuel cell is a galvanic cell that requires a constant external supply of one or more reactants in order to generate electricity because the products of the reaction are continuously removed from the cell. Perhaps the most widely known battery in the mass market is the alkaline battery, which contains an electrolyte in an acidic water-based paste and is adapted to operate under alkaline conditions. There are also several other types of batteries: lithium-iodine, which consist of a solid electrolyte; nickel-cadmium, which are rechargeable; and lead acid, which are also rechargeable and find extensive use as car batteries.

19.6. Corrosion

Corrosion is the galvanic process by which metals deteriorate through oxidation. Corrosion can be protected against by adding a second layer of a metal that is harder to oxidize. Galvanized steel, for example, is protected by a thin layer of zinc. Corrosion can also be protected against by adding a more easily oxidized metal to the surface. Sacrificial electrodes would be an example of this process.

19.7. Electrolysis

Electrolysis uses an external voltage to drive a nonspontaneous reaction to completion. Electrolysis is useful for such things as producing hydrogen and oxygen gases from water, producing sodium metal from salt mixtures, and producing aluminum. The reaction can be driven even further by the application of an extra voltage known as an *overvoltage*. Electroplating is the process by which a second metal is deposited over a metal surface and is useful for corrosion prevention and for enhancing the metal's value. The amount of material consumed or produced in an electrolysis reaction can be calculated from the reaction stoichiometry, the current passed, and the duration of the reaction.

Self-Test

1. An electrochemical cell that generates electricity from a spontaneous reaction is a(n) _____
   A. electrolytic cell.
   B. fuel cell.
   C. galvanic cell.
   D. concentration cell.

2. Write the reaction for the oxidation of copper metal to copper(II) ions.
   A. $Cu(s) \rightarrow Cu^{2+}(aq) + 2e^-$
   B. $Cu^{2+}(aq) + 2e^- \rightarrow Cu(s)$
   C. $Cu(s) + 2e^- \rightarrow Cu^{2+}(aq)$
   D. $Cu(s) \rightarrow 2Cu^+(aq) + 2e^-$

3. The reductive half-reaction occurs in what part of an electrochemical cell?
   A. Anode
   B. Cathode
   C. Salt bridge
   D. Voltmeter

4. Write the balanced cell reaction for the following:
$$Ce^{4+}(aq) + I^-(aq) \rightarrow I_2(s) + Ce^{3+}(aq)$$
   A. $Ce^{4+}(aq) + I^-(aq) + e^- \rightarrow I_2(s) + Ce^{3+}(aq) + 2e^-$
   B. $2I^-(aq) + 2Ce^{4+} \rightarrow 2Ce^{3+}(aq) + I_2(s)$
   C. $Ce^{4+}(aq) + I^-(aq) + 2e^- \rightarrow I_2(s) + Ce^{3+}(aq) + e^-$
   D. $2I^-(aq) \rightarrow I_2(s) + 2e^-$

5. Using the data in Table 19.2 of the text, calculate the standard cell potential for the reaction in Question 4.

   A. 1.18 V

   B. 2.26 V

   C. 2.9 V

   D. 3.98 V

6. Which of the following is not a common reference electrode?

   A. Saturated calomel electrode

   B. Standard hydrogen electrode

   C. Glass electrode

   D. Nernst electrode

7. Which of the following is not a standard condition for the standard hydrogen electrode?

   A. 1 M solutions

   B. 25°C

   C. Ambient pressure gases

   D. 0 V

8. Which of the following would you expect to be the strongest oxidizer?

   A. Co^{2+}

   B. Cl_2

   C. Ce^{4+}

   D. In^{3+}

9. Which of the following would you expect to be the strongest reducer?

   A. H^+

   B. Ni^{2+}

   C. Fe^{2+}

   D. Li^+

10. What must be the value of $[Fe^{2+}]/[Fe^{3+}]$ in the cell $Pt|Fe^{3+}(aq)$, $Fe^{2+}(aq)||Hg_2^{2+}(aq, 1.0$ mol/L$)|Hg$ for it to have a potential of 0.060 V?

   A. 4.3

   B. 4.7

   C. 3.7

   D. 3.5

11. Calculate the potential of the following cell:

$$Zn|Zn^{2+}(aq, 0.1 \text{ mol/L})||Ni^{2+}(aq, 0.0010 \text{ mol/L})|Ni$$

A. 0.47 V
B. 0.18 V
C. -1.45 V
D. 0.62 V

12. Calculate the reaction quotient, Q, for the following cell:

$$Pt|Sn^{4+}(aq), Sn^{2+}(aq)||Pb^{4+}(aq), Pb^{2+}(aq)|C(gr), E = 1.33 \text{ V}$$

A. 1.5×10^6
B. 2.1×10^6
C. 3×10^6
D. 4.2×10^6

13. Determine the unknown in the following cell:

$$Pt|H_2(g, 1.0 \text{ atm})|H^+(pH = ?)||Cl^-(aq, 1.0 \text{ mol/L})|Hg_2Cl_2(s)|Hg, E = 0.33 \text{ V}$$

A. 1.0
B. 1.7
C. 2.3
D. 2.9

14. Which of the following is not a type of battery?
A. Nickel-cadmium
B. Fuel cell
C. Lead-acid
D. Lithium-iodine

15. The typical output of an alkaline battery is _____
A. 3.5 V.
B. 2.04 V.
C. 1.4 V.
D. 1.5 V.

16. The type of battery most likely to be used in a car is a(n) _____
A. lithium-iodine.
B. lead-acid.
C. nickel-cadmium.
D. alkaline.

17. One method of protecting against corrosion is _____
    A. overvoltage.
    B. the Hall-Heroult process.
    C. galvanization.
    D. reduction.

18. Corrosion is what kind of reaction?
    A. Oxidation
    B. Reduction

19. Which of the following is not used to calculate the amount of material produced or consumed by an electrolysis reaction?
    A. Reaction stoichiometry
    B. Strength of oxidizer
    C. Current passed
    D. Duration of reaction

20. An aqueous solution of Mn^{2+} is electrolyzed. Which substance will be reduced at the cathode?
    A. Mn^{2+}
    B. H_2O

Answers: 1. C; 2. A; 3. B; 4. B; 5. D; 6. D; 7. C; 8. C; 9. D; 10. B; 11. A; 12. C; 13. A; 14. B; 15. D; 16. B; 17. C; 18. A; 19. B; 20. A

Nuclear Chemistry

Key Words

alpha decay
α particle
beta decay
breeder reactor
cosmic rays
cosmogenic radiation
critical mass
daughter nuclei
dose
electron capture
gamma emission
gray
half-life
heavy-water reactor
ionizing radiation
light-water reactor
mass defect

medical imaging
nonionizing radiation
nuclear binding energy
nuclear chain reaction
nuclear decay reaction
nuclear fission
nuclear fusion
nuclear fusion reactor
nuclear transmutation
   reaction
nucleons
nuclide
parent nucleus
positron
positron emission
positron emission
   tomography

rad
radiation therapy
radioactive
radioactive decay
radioactive decay series
radioactivity
radioisotopes
radon
rem
roentgen
spontaneous fission
supercritical mass
super-heavy elements
terrestrial radiation
thermonuclear reaction
transuranium elements

By the end of this chapter, you should be able to:

- Understand the factors that affect nuclear stability
- List the different kinds of radioactive decay
- Write a balanced nuclear reaction equation
- Understand a radioactive decay series
- Distinguish between ionizing and nonionizing radiation and their effects on matter
- Name natural and artificial sources of radiation
- Calculate a mass-energy balance
- Calculate a nuclear binding energy
- Distinguish between nuclear fission and fusion
- Understand how different types of nuclear reactors operate
- Understand how elements were formed

Chapter Overview

In previous chapters, we've spent a good deal of time talking about various kinds of chemical reactions. In this chapter, we will discuss a different kind of reaction: a nuclear reaction. Where chemical reactions are restricted to the electrons of an atom, a nuclear reaction involves the nucleus of an atom, and nuclear reactions change the identity of the element(s) involved and release exponentially more energy than any chemical reaction.

20.1. The Components of the Nucleus

A radioactive element has an unstable nucleus that decays spontaneously and whose emissions are collectively called *radioactivity*. Elemental isotopes that emit radiation are called *radioisotopes*. The protons and neutrons that make up an atom's nucleus are referred to as *nucleons*. A nuclide is an atom with a particular number of protons and neutrons. Stable atomic nuclei generally have even numbers of both protons and neutrons and a proton-to-neutron ratio of at least 1.

20.2. Nuclear Reactions

In a nuclear decay reaction, the parent nucleus is converted to a more stable daughter nucleus. Nuclei that have too few neutrons will decay by converting a proton to a neutron in positron emission, while nuclei that have too many neutrons will decay by converting a neutron to a proton in beta decay. Nuclei with $Z > 83$ are unstable and generally decay by emitting an α particle in alpha decay. Radioactive decay conserves both the total positive charge and the total number of nucleons. An α particle is denoted as $_2^4\alpha$, and alpha decay results in a daughter nucleus with a mass number lowered by 4 and an atomic number lowered by 2 from the parent nucleus.

Beta decay results in a daughter nucleus with the same mass number and an atomic number that is increased by 1 over the parent nucleus. Positron emission results in a daughter nucleus with the same mass number and an atomic number that is decreased by 1 over the parent nucleus. In electron capture, an inner shell electron reacts with a proton to produce a neutron, with the accompanying emission of an X ray. This results in a daughter nucleus with the same mass number and an atomic number 1 lower than the parent. Gamma emission occurs when an excited daughter nucleus decays to a lower energy state, resulting in the emission of a γ ray. Very heavy nuclei can undergo spontaneous fission and break into two pieces that can have very different atomic numbers and masses. This can occur through a radioactive decay series, a succession of alternating alpha- and beta-decay reactions. All the transuranium elements are artificial and are synthesized through nuclear transmutation reactions, where the target nucleus is bombarded with energetic subatomic particles to result in a larger product nucleus.

20.3. The Interaction of Nuclear Radiation with Matter

The energy of radiation determines its effect on matter. Nonionizing radiation is relatively low in energy, and the energy it transfers to matter takes the form of heat. Ionizing radiation, on the other hand, is higher in energy. When it collides with an atom, it has the energy to completely remove an electron to form a positively charged ion that can cause damage to biological tissue. Alpha particles do not penetrate far into matter, whereas γ rays penetrate much more deeply. The common units of radiation exposure are the roentgen (R), the amount of energy absorbed by dry air, and the rad, the amount of radiation that produces 0.01 J of energy in 1 kg of matter. The rem is a unit of measure for the actual amount of tissue damage caused by a given amount of radiation. There are several kinds of naturally occurring radiation: cosmic radiation, which consists of high-energy particles and γ rays that are emitted by the sun and other stars; cosmogenic radiation, which is produced by the interaction of cosmic rays with gases in the upper atmosphere; and terrestrial radiation, which comes from radioactive elements that were present in the primordial earth and their decay products.

20.4. Thermodynamic Stability of the Atomic Nucleus

Nuclear reactions generate large changes in energy that result in measurable changes in mass. The change in mass is related to the change in energy by Einstein's equation, $\Delta E = (\Delta m)c^2$, with large changes in energy reported in units of keV or MeV (thousands of millions of electronvolts). The mass defect of the nucleus is the difference between the experimentally determined mass of an atom and the sum of the masses of the component protons, neutrons, and electrons. The mass defect corresponds to a nuclear binding energy, the amount of energy released when a nucleus forms from its component particles. In the process of nuclear fission, nuclei are split into lighter nuclei with an accompanying release of several neutrons and large amounts of energy. Nuclear fusion is a process in which two lighter nuclei combine to produce a heavier nuclei and a large amount of energy. The critical mass is the minimum mass required to support a self-sustaining nuclear chain reaction.

20.5. Applied Nuclear Chemistry

Nuclear power plants use nuclear reactions to generate electricity. There are several types of nuclear reactors, including light-water reactors, heavy-water reactors, breeder reactors, and nuclear-fusion reactors. Light-water reactors use enriched uranium as fuel. Heavy-water reactors can use unenriched uranium as fuel because heavy water shields from neutrons more effectively than regular water. A breeder reactor produces more fissionable fuel than it consumes. A nuclear fusion reactor requires extremely high temperatures in order to drive the reactions. Thermonuclear reactions require very high temperatures in order to initiate the reaction.

20.6. The Origin of the Elements

Hydrogen is the most abundant element in the universe. Stars such as the sun fuel themselves by fusing hydrogen nuclei into helium nuclei. Heavier elements are formed in the interior of stars by the fusion of helium and hydrogen nuclei. Successive fusions may form elements up to magnesium. Elements as high as iron-56 and nickel-58 can also be formed by similar processes carried out at higher temperatures. Heavier elements than these can only be formed during the explosion of a supernova and involve a process of multiple neutron-capture events.

Self-Test

1. Which of the following is not an isotope of oxygen?
   A. Oxygen-15
   B. Oxygen-16
   C. Oxygen-17
   D. Oxygen-18

2. If the half-life of a radioactive element is 8 years, what percentage of the original sample would be left after 64 years?
   A. 2.25%
   B. 1.56%
   C. 0.39%
   D. 0.064%

3. $^{222}_{86}Rn$ decays to $^{218}_{84}Po$ through what process?
   A. Beta decay
   B. Alpha decay
   C. Positron emission
   D. Gamma emission

4. $^{56}_{26}Fe$ decays to $^{56}_{25}Mn$ through what process?
   A. Beta decay
   B. Alpha decay
   C. Positron emission
   D. Gamma emission

5. Which of the following processes is accompanied by the emission of an X ray?
   A. Positron emission
   B. Electron capture
   C. Gamma emission
   D. Spontaneous fission

6. The most energetic form of radioactive decay is _____
   A. positron emission.
   B. electron capture.
   C. gamma emission.
   D. spontaneous fission.

7. Which of the following is not a natural source of radiation?
   A. Cosmogenic radiation
   B. Ionizing radiation
   C. Terrestrial radiation
   D. Cosmic radiation

8. The largest source of radiation exposure for a person is _____

   A. internal radiation.

   B. cosmogenic radiation.

   C. radon.

   D. X rays.

9. The unit used to describe the actual damage done to tissue by a given amount of radiation is the _____

   A. roentgen.

   B. rad.

   C. gray.

   D. rem.

10. Calculate the change in energy that accompanies the decay of ^{238}U to ^{234}Th (ignore other particles).

    A. 1.7×10^{17} J/mol

    B. 3.6×10^{17} J/mol

    C. 2.4×10^{17} J/mol

    D. 2.9×10^{17} J/mol

11. The minimum amount of material needed to support a self-sustaining nuclear reaction is the _____

    A. mass defect.

    B. nuclear binding energy.

    C. critical mass.

    D. strong nuclear force.

12. Which type of nuclear reactor requires the highest temperatures?

    A. Light-water reactor

    B. Heavy-water reactor

    C. Breeder reactor

    D. Fusion reactor

13. Which type of reactor uses enriched uranium as a fuel?

    A. Light-water reactor

    B. Heavy-water reactor

    C. Breeder reactor

    D. Fusion reactor

14. Hydrogen is fused into helium during the _____ stage of a star's lifetime.

    A. yellow star

    B. red giant

    C. super red giant

    D. supernova

15. The heaviest elements are synthesized during the _____ stage of a star's lifetime.
    A. yellow star
    B. red giant
    C. super red giant
    D. supernova

Answers: 1. A; 2. C; 3. B; 4. C; 5. B; 6. D; 7. B; 8. C; 9. D; 10. B; 11. C; 12. D; 13. A; 14. A; 15. D

CHAPTER 21

Periodic Trends and the *s*-Block Elements

Key Words

crown ethers

cryptands

deuterium

effective nuclear charge

graphite intercalation
  compounds

hydride ion

hydrogen bond

inert-pair effect

ionophores

ion pumps

organometallic compounds

protium

three-center bond

tritium

By the end of this chapter, you should be able to:

- Describe the physical and chemical properties of hydrogen
- Be familiar with the reactions, compounds, and complexes of the alkali and alkaline earth metals
- Describe processes by which the alkali and alkaline earth metals are isolated
- List uses of the alkali and alkaline earth metals
- Understand the biological role that *s*-block elements play

Chapter Overview

In the next few chapters, we'll make our way through the periodic table and discuss trends, properties, and reactions of the various groups of elements. This chapter begins by discussing the *s*-block elements: the alkali and alkaline earth metals, and hydrogen.

21.1. Overview of Periodic Trends

The systematic increase in atomic number and the filling of electron orbitals give rise to definite periodic trends in atomic properties. The most fundamental property that leads to periodic variations is the effective nuclear charge (Z_{eff}). The mixture of metals, nonmetals, and semimetals in Groups 13, 14, and 15 can make the chemistry of those groups complicated. Increasing the ionization energies and decreasing the bond lengths lead to the inert-pair effect, which causes the heaviest elements of Groups 1 and 2 to have a stable oxidation state that is lower by 2 than the maximum predicted for their respective groups.

21.2. The Chemistry of Hydrogen

There are three isotopes of hydrogen: protium, deuterium, and tritium, all of which have different properties. Deuterium and tritium are tracers, used by biochemists to follow the path of a molecule through an organism or a cell. Hydrogen can form compounds in several ways: through a proton, a hydride ion, an electron-pair bond, a hydrogen bond, or a three-center (or electron-deficient) bond. Hydrogen gas is generated by the reaction of an active metal with dilute acid, aluminum or zinc with a strong base, or on an industrial scale by catalytic steam reforming.

21.3. The Alkali Metals

Sodium and potassium were the first alkali metals to be isolated by passing an electric current through molten potassium and sodium carbonates. The alkali metals are among the strongest reductants known. They can be isolated from their molten salts through electrolysis, recovered from silicate ores through a lengthy process, or separated from their hydroxide salts. The alkali metals react with the halogens to form ionic halides. They react with the heavier chalcogens to produce metal chalcogenides and with oxygen to form compounds whose stoichiometry depends on the size of the metal. The heavier alkali metals react with graphite to form graphite intercalation compounds, where metal atoms are inserted between the sheets of carbon atoms. Alkali metals react with the heavier Group 14 elements to give polyatomic anions with three-dimensional cage structures. All of the alkali metals react with hydrogen at high temperatures to produce the corresponding hydrides, and they also reduce water to produce hydrogen gas. Alkali metal salts can be prepared by the reaction of the metal hydroxide with an acid, followed by the evaporation of water. Alkali metals can also react with acidic organic compounds to produce salts. In addition, they can form organometallic compounds, which have properties that differ from those of both their metallic and organic components.

21.4. The Alkaline Earth Metals

Most of the alkaline earth metals can be isolated by electrolyzing their chlorides or oxides. The alkaline earth metals have little to no affinity for an added electron. All alkaline earth metals react with the halogens to form halides. Barium reacts with oxygen to form a peroxide, whereas the other alkaline earth metals will react with oxygen to form oxides. They react with the heavier chalcogens to form chalcogenides or polychalcogenide ions. All the alkaline earth metals except beryllium react with carbon and hydrogen to form carbides and hydrides. All of them react with nitrogen gas to form nitrides. The alkaline earths have a tendency to complex with crown ethers, cryptands, and other Lewis bases. The most important alkaline earth organometallic compounds are the Grignard reagents, which are used to synthesize organic compounds.

21.5. The *s*-Block Elements in Biology

Covalent hydrides in which a hydrogen is bonded to oxygen, nitrogen, or sulfur are polar, hydrophilic molecules that form hydrogen bonds and undergo acid-base reactions. These hydrogen-bonding interactions are critical in stabilizing the structure of proteins and DNA. Hydrogen-bonding interactions are also responsible for the unique properties of water that allow life to exist on this planet. Group 1 and Group 2 metals are present in organisms in macrominerals, which are important components of intracellular and extracellular fluids. Ion pumps bind ions based on their charge and radius and selectively transport metal ions across cell membranes. Ionophores work to facilitate the transport of metal ions across membranes.

Self-Test

1. The properties of Li are most similar to the properties of _____
   A. H.
   B. Be.
   C. Mg.
   D. Na.

2. Elements with the highest ionization energy will be found in what part of the periodic table?
   A. Upper left
   B. Upper right
   C. Lower left
   D. Lower right

3. Elements with the lowest effective nuclear charge will be found in what part of the periodic table?
   A. Upper left
   B. Upper right
   C. Lower left
   D. Lower right

4. Which of the following is not an isotope of hydrogen?

    A. Hydridium

    B. Protium

    C. Tritium

    D. Deuterium

5. Which of the following is not a way in which hydrogen can form a compound?

    A. Hydrogen bond

    B. Three-center bond

    C. Ionic bond

    D. Complexation

6. The starting material in catalytic steam reforming is _____

    A. ionic hydrides.

    B. covalent hydrides.

    C. hydrocarbons.

    D. metallic hydrides.

7. Which of the alkali metals reacts with atmospheric nitrogen?

    A. K

    B. Cs

    C. Na

    D. Li

8. Which of the alkali metal salts are used as a drying agent?

    A. Na

    B. K

    C. Rb

    D. Cs

9. Which of the alkali metals forms a stable oxide and nitride?

    A. Li

    B. Na

    C. K

    D. Rb

10. Which of the alkaline earth metals does not form a carbide when reacted with carbon?

    A. Mg

    B. Be

    C. Ca

    D. Ba

11. Which of the alkaline earth metals is highly radioactive?

    A. Mg

    B. Ra

    C. Ba

    D. Sr

12. Which of the alkaline earth metals would you expect to form the most soluble carbonate salt?

    A. Be

    B. Mg

    C. Ca

    D. Sr

13. What gives water its high heat capacity?

    A. The electronegativity of oxygen

    B. The covalent bonds between the O and H

    C. Hydrogen bonding

    D. The bent structure of the molecule

14. Which element has a link to hypertension?

    A. Ca

    B. Na

    C. Mg

    D. K

15. Metal ions are transported across cell membranes by _____

    A. crown ethers.

    B. macrominerals.

    C. cryptands.

    D. ion pumps.

Answers: 1. C; 2. B; 3. C; 4. A; 5. D; 6. C; 7. D; 8. A; 9. A; 10. B; 11. B; 12. A; 13. C; 14. C; 15. D

CHAPTER 22

The *p*-Block Elements

Key Words

chalcogens
halogens
noble gases
pnicogens

By the end of this chapter, you should be able to:

- Understand how periodic trends and electron configurations determine the chemical properties of the *p*-block elements
- Use kinetics and thermodynamics to understand the reactivity of the *p*-block elements
- Understand the trends down the groups of *p*-block elements
- Predict the types of reactions that *p*-block elements undergo

Chapter Overview

In the last chapter, we began our discussion of the chemistry of the elements by focusing on the *s* block. In this chapter, we will continue the discussion with the elements of the *p* block.

22.1. The Elements of Group 13

Compounds of Group 13 elements are thermodynamically stable, making the elements difficult to isolate. Boron behaves like a nonmetal, while the heavier elements behave like metals. The properties of Group 13 elements display some inconsistencies that can be attributed to poor shielding that leads to an increase in Z_{eff}. Boron has unique properties in that it does not form a metal lattice; instead it forms aggregates with multicenter bonds, including metal borides, where boron is bonded to other boron atoms to form three-dimensional clusters with geometric structures. Neutral Group 13 compounds behave like Lewis acids. Group 13 oxides dissolve in dilute acids. The stability of the Group 13 hydrides decreases as you go down the group.

22.2. The Elements of Group 14

The elements of Group 14 show the greatest range of behavior of any group in the periodic table. The stability of the $+2$ oxidation state increases from carbon to lead. The tendency to form multiple bonds decreases as atomic number increases. Carbon forms three kinds of carbides with less electronegative elements: covalent carbides, whose three-dimensional structures make them hard, high melting, and chemically inert; interstitial carbides, whose covalent metal–carbon interactions make them among the hardest substances known; and ionic carbides, which contain metal cations and methide or acetylide anions. Metallic behavior increases as you move down the group. Dioxides of the group elements become more basic and more metallic as you move down the group. Group 14 hydrides become thermodynamically less stable as you go down the group. As atomic size increases, multiple bonds involving the group elements become weaker due to poor overlap and steric hindrance.

22.3. The Elements of Group 15 (the Pnicogens)

Among the Group 15 elements, nitrogen and phosphorus behave like nonmetals, arsenic and antimony behave like semimetals, and bismuth behaves like a metal. Nitrogen is capable of forming compounds in nine different oxidation states. Neutral covalent compounds of the pnicogens are classified as Lewis bases. Nitrogen forms stable multiple bonds with other second-period elements. Nitrogen can also react with electropositive elements to produce solids that range from covalent to ionic in character. Reactions with electropositive metals will produce ionic nitrides, reactions with less electropositive metals will produce interstitial nitrides, and reactions with semimetals will produce covalent nitrides. Reactions of phosphorus with metals produce phosphides. The reactivity of the pnicogens decreases with increasing atomic number.

22.4. The Elements of Group 16 (the Chalcogens)

The lightest member of the chalcogens, oxygen, has the greatest tendency to form multiple bonds. Oxygen has a high electronegativity; consequently, it forms compounds in which it has a negative oxidation state and highly polar bonds. Nonmetal oxides have acidic character, whereas metal oxides have basic character. Semimetal oxides are amphoteric. The electronegativity of the chalcogens decreases as you move down the group and tends to form compounds in the -2 oxidation state.

22.5. The Elements of Group 17 (the Halogens)

With the exception of iodine, all of the halogens are found in nature as halide salts. They are nonmetals and are too reactive to exist in nature as free elements. Fluorine is the most reactive element in the periodic table. Compounds of fluorine and electropositive elements are ionic, whereas compounds of fluorine and less-electropositive elements and metals are covalent. All of the halogens react with hydrogen to produce hydrogen halides. Halogens, with the exception of fluorine, are also capable of forming oxoacids in water as well as interhalogen compounds.

22.6. The Elements of Group 18 (the Noble Gases)

The noble gases have a closed-shell valence-electron configuration. Highly electronegative elements are the only ones that can form stable compounds with the noble gases without being oxidized themselves. The ionization energies of the noble gases decrease with increasing atomic number.

Self-Test

1. Group 13 elements are _____ found in nature in their free state.
   A. always
   B. sometimes
   C. rarely
   D. never

2. Which of the Group 13 elements behaves like a nonmetal?
   A. Boron
   B. Aluminum
   C. Gallium
   D. Indium

3. Which of the following is not an allotrope of carbon?
   A. Diamond
   B. Graphite
   C. Carbides
   D. Fullerenes

4. Which of the following would you expect to be more stable?
   A. CCl_2
   B. $SiCl_2$
   C. $SnCl_2$
   D. $PbCl_2$

5. Nitrogen is capable of forming compounds in how many different oxidation states?
   A. 5
   B. 6
   C. 7
   D. 9

6. Nitrogen reacts with _____ to form covalent nitrides.
   A. electropositive metals
   B. electropositive nonmetals
   C. semimetals
   D. less electropositive metals

7. Which of the following is not a property of a metal-rich phosphide?
   A. Low melting point
   B. Hard
   C. Electrically conductive
   D. Metallic luster

8. Which of the following would you expect to have the lowest reactivity?
   A. S
   B. Te
   C. Po
   D. Se

9. Metal oxides are usually _____
   A. acidic.
   B. basic.
   C. amphoteric.

10. Which of the following would you expect to have the highest electronegativity?
    A. Se
    B. Po
    C. O
    D. S

11. Which of the following is not found in nature as a halide salt?

 A. F

 B. Cl

 C. Br

 D. I

12. Which of the following does not react with water to form an oxoacid?

 A. F

 B. Cl

 C. Br

 D. I

13. Which of the following is the most reactive element in the periodic table?

 A. Cl

 B. F

 C. Br

 D. I

14. Which of the following would you expect to have the highest ionization energy?

 A. Ne

 B. Ar

 C. Kr

 D. Xe

15. Xenon can form compounds with oxidation states as high as _____

 A. 5.

 B. 6.

 C. 7.

 D. 8.

Answers: 1. D; 2. A; 3. C; 4. D; 5. D; 6. C; 7. A; 8. C; 9. B; 10. C; 11. D; 12. A; 13. B; 14. A; 15. D

CHAPTER 23

The *d*-Block Elements

Key Words

chelate effect	hard acid	metalloprotein
complex ions	hard base	siderophores
coordination compound	isomers	soft acid
crystal field splitting energy	Jahn-Teller effect	soft base
crystal field theory	ligands	spin-pairing energy
geometrical isomers	metal complex	structural isomers
	metalloenzyme	

By the end of this chapter, you should be able to:

- Know the trends in the *d*-block elements
- Understand the chemistry of the transition metals
- Understand how metals are extracted from ores
- Recognize the most common structures for metal complexes
- Predict the relative stability of metal complexes
- Have a basic understanding of crystal field theory
- Be familiar with the roles of transition-metal complexes in biological systems

Chapter Overview

In the previous chapters, we've discussed the *s*-block and *p*-block elements of the periodic table. In this chapter, we finish our discussion of the periodic table by focusing on the *d*-block elements.

23.1. General Trends among the Transition Metals

The transition metals are characterized by partially filled *d* subshells. Several irregularities in the filling order of the subshells can arise among the elements of this block, and they can be difficult to predict due to the similar energies of the *ns, nd,* and *nf* orbitals. There is a greater increase in atomic radius between the 3*d* and 4*d* metals than between the 4*d* and 5*d* metals due to the lanthanide contraction. Generally speaking, as you go across a row in the periodic table, the ionization energies, electronegativities, densities, thermal conductivities, and electrical conductivities increase, while the enthalpies of hydration decrease. Any anomalies in the patterns can be explained by the increasing stability of partially filled and filled subshells. Many transition metals can form compounds in several oxidation states, with the higher oxidation states becoming less stable across the row and more stable down the column.

23.2. A Brief Survey of Transition Metal Chemistry

The transition elements of Group 3 are highly electropositive and are powerful reductants. They have a high affinity for oxygen, as do the elements of Group 4 and Group 5. The hydrides, nitrides, carbides, and borides of Group 4 and Group 5 are hard, high-melting-point solids that conduct electricity. Group 6 elements are less electropositive and have a maximum oxidation state of +6, making their compounds covalent in character. The Group 6 elements also form nitrides, carbides, and borides with properties similar to those of Groups 4 and 5. Group 7 elements have a maximum oxidation of +7. The nitrides, carbides, and borides of Group 7 are stable at high temperatures and have metallic properties. Elements of Groups 8, 9, and 10 have such high ionization potentials that the complete loss of valence electrons occurs rarely, if at all. These elements form a range of binary nitrides, carbides, and borides. The Group 11 metals are known as the coinage metals and are relatively unreactive but can form halide compounds. The Group 12 elements are found in nature compounded with sulfur. Mercury is the only metallic element that is liquid at room temperature.

23.3. Metallurgy

Metallurgy is the process by which metals are extracted from their ores. It consists of three basic steps: mining, separation and concentration, and reduction. There are different types of separation and reduction methods. Setting and flotation are methods that separate based upon differences in density. Pyrometallurgy is a chemical reduction carried out at elevated temperatures in the presence of a reductant that does not form a stable compound with the metal being reduced. Hydrometallurgy carries out a chemical or electrochemical reduction in an aqueous solution via the formation of metal complexes.

23.4. Coordination Compounds

Metal complexes are polyatomic species in which a transition metal ion is bound to one or more ligands. Complex ions are electrically charged metal complexes. Coordination compounds contain one or more metal complexes. The coordination number of the complex determines the structure it will have. Complexes with a tetrahedral or square planar shape have a coordination number of 4. Trigonal bipyramidal and square pyramidal complexes have a coordination number of 5. Octahedral and trigonal prisms have a coordination number of 6. Larger metal ions may have coordination numbers of 7, 8, or 9, with multiple structures possible for these complexes. Lewis bases can either be hard bases, with small, relatively nonpolarizable donor atoms, or soft bases, with larger and relatively polarizable donors. Hard acids have a high affinity for hard bases, whereas soft acids have a high affinity for soft bases. Many metal complexes exist as isomers, which are two or more compounds that have the same formula but different arrangements of the atoms.

23.5. Crystal Field Theory

Crystal field theory is a bonding model that is used to explain properties of transition metals that cannot be explained by valence bond theory. The theory says that complexes arise from electrostatic interactions between the central metal ion and negatively charged ligands or dipoles around the metal ion. The arrangement of the ligands has the potential to split the d orbitals into sets of orbitals with different energies. The *crystal field splitting energy* (Δ_0) is defined as the difference between the energy levels in an octahedral complex and depends upon the charge of the metal ion, its position in the periodic table, and the ligands involved in the complex. The spin-pairing energy is the increase in energy that occurs when an electron is added to an already occupied orbital. A high-spin configuration results when the crystal field splitting energy is less than the spin-pairing energy, while a low-spin configuration results when the crystal field splitting energy is greater than the spin-pairing energy. A large Δ_0 results when a strong-field ligand interacts strongly with the d orbitals, whereas a smaller Δ_0 results when a weak-field ligand interacts weakly with the d orbitals. The crystal field stabilization energy is the additional stabilization that arises from placing electrons in a lower-energy set of d orbitals.

23.6. Transition Metals in Biology

There are a number of transition metals that are essential for living organisms. In order to be useful, the metals must be moved from the surrounding environment into the organism and into the cells of the organism. Iron is one such essential metal. Bacteria use organic ligands that have a high affinity for iron and increase the concentration of dissolved iron available to the organism. Mammals dissolve iron in the highly acidic environment of their stomachs. The iron is absorbed in the small intestine and transported to where it is needed. Metalloproteins contain one or more metal ions tightly bound within the protein. They may catalyze biochemical reactions or transfer electrons from one place to another.

Self-Test

1. In what part of the *d* block would you expect the electronegativity to be the greatest?

   A. Upper left

   B. Upper right

   C. Lower left

   D. Lower right

2. The highest possible oxidation state would be the least stable in elements of _____

   A. Group 3.

   B. Group 5.

   C. Group 7.

   D. Group 9.

3. Which metal is liquid at room temperature?

   A. Chromium

   B. Molybdenum

   C. Mercury

   D. Tungsten

4. The chemistry of the Group 6 metals is dominated by the _____ oxidation state.

   A. +3

   B. +4

   C. +5

   D. +6

5. The coinage metals belong to which Group?

   A. Group 12

   B. Group 11

   C. Group 10

   D. Group 9

6. Which of the following is not a general step in metallurgy?

   A. Separating

   B. Reducing

   C. Smelting

   D. Mining

7. Iron is refined in a _____
    A. blast furnace.
    B. leeching tank.
    C. sifter.
    D. strip mine.

8. The most common shape for a metal complex with a coordination number of 6 is _____
    A. tetrahedral.
    B. trigonal bipyramidal.
    C. octahedral.
    D. trigonal prism.

9. Hard acids prefer hard bases _____
    A. always.
    B. most of the time.
    C. some of the time.
    D. never.

10. Which of the following is not a type of isomer found in metal complexes?
    A. Structural isomers
    B. *Cis* isomers
    C. *Trans* isomers
    D. Stereoisomers

11. The crystal field splitting energy depends upon all of the following except _____
    A. the location of the metal in the periodic table.
    B. the spin-pairing energy.
    C. the charge on the metal ion.
    D. the ligands.

12. The crystal field stabilization energy explains anomalies observed in _____
    A. electron configuration.
    B. d^9 complexes.
    C. lattice energies.
    D. compound color.

13. In mammals, iron is absorbed into the bloodstream through the _____
    A. skin.
    B. stomach.
    C. intestines.
    D. lungs.

14. Oxygen is bound and transferred by _____
    A. siderophores.
    B. transferring.
    C. cytochrome-c.
    D. hemoglobin.

15. Siderophores are responsible for the uptake of _____
    A. iron.
    B. copper.
    C. nickel.
    D. mercury.

Answers: 1. B; 2. D; 3. C; 4. D; 5. B; 6. C; 7. A; 8. C; 9. A; 10. D; 11. B; 12. C; 13. C; 14. D; 15. A

CHAPTER 24

Organic Compounds

Key Words

addition reaction
carbanion
carbocation
carbohydrates
conformational isomer
electrophile
elimination reaction
free radical
functional groups

Grignard reagent
isomers
lipids
nucleic acids
nucleophile
nucleophilic substitution
  reaction
organic compounds
proteins

pyrolysis reaction
quaternary ammonium
  salt
specific rotation
stereoisomer
structural isomer
substitution reaction

By the end of this chapter, you should be able to:

- Understand the different types of isomers
- Be familiar with the most common kinds of organic reactions
- Be familiar with the most common organic functional groups
- Identify common types of biologically relevant molecules

Chapter Overview

We close out this textbook by revisiting organic chemistry, which we have discussed somewhat throughout the course of this text. In this chapter, we address it in more detail. We discuss a number of classes and common functional groups in organic chemistry, as well as the concept of isomerism. We conclude the chapter with a brief presentation of the major classes of biological organic compounds.

24.1. Functional Groups and Classes of Organic Compounds

Organic compounds can be classified into several major categories, depending on the types of functional groups that they contain. A functional group is a structural unit that determines the chemical reactivity of a molecule under a given set of conditions. Many organic compounds have common names and systematic names. The systematic names are based on a foundation that denotes the hydrocarbon framework, with numbers indicating the carbon on which a particular functional group is located.

24.2. Isomeric Variations in Structure

There are several types of isomers in organic structures. Rotation about a σ bond can produce three-dimensional structures known as *conformational isomers*. Structural isomers have the same molecular formula but different connectivities of the atoms. Stereoisomers occur when molecules have the same connectivities but different orientations in space. Stereoisomers can be further broken down into geometric isomers, which differ in the placement of substituents in a rigid molecule, or optical isomers, which are nonsuperimposable mirror images. A set of molecules that are not superimposable mirror images of each other are said to be *chiral*. A molecule and its nonsuperimposable image are called *enantiomers*. Enantiomers differ in their interaction with plane polarized light. A compound is said to be optically active if its solution rotates the plane polarized light in one direction, and optically inactive if it produces rotations that cancel each other out. Enantiomers can be separated into *d* forms, which rotate light clockwise, and *l* forms, which rotate the light counterclockwise. A solution containing equal concentrations of *d* and *l* enantiomers is called a *racemic mixture*.

24.3. Reactivity of Organic Molecules

The reactivity of an organic molecule can be affected by the degree of substitution of the carbon bonded to a functional group, with a primary carbon having one other group bonded to it, a secondary carbon having two groups bonded, and a tertiary carbon having three other groups bonded to it. Reactions involving organic molecules will often produce intermediate products, which can be useful in determining the pathway of the reaction. A carbocation is an intermediate with six valence electrons. A free radical is an intermediate that is electron-deficient but neutral. A carbanion contains eight valence electrons and is negatively charged.

24.4. Common Classes of Organic Reactions

There are several different types of organic reactions that can occur. In a substitution reaction, one atom or group of atoms in a molecule is replaced by another atom or group of atoms from another substance. In an elimination reaction, adjacent atoms are removed, resulting in the formation of a multiple bond and a small molecule. The reverse of an elimination reaction is an addition reaction. Organic molecules can also undergo free-radical reactions and oxidation-reduction reactions.

24.5. General Properties and Reactivity of Functional Groups

The functional groups in organic molecules play a large role in determining the physical properties and reactivity of the molecule. Alkanes are saturated hydrocarbons that can undergo such reactions as catalytic cracking, in which high temperatures are used to cleave weak bonds and convert straight-chained molecules to highly branched ones. The double bond of an alkene gives rise to the possibility of *cis* and *trans* isomers. An alcohol undergoes reactions that cleave either the O—H bond or the C—O bond. Phenols are acidic, aromatic alcohols. Ethers are relatively unreactive. Aldehydes and ketones are generally prepared from the oxidation of alcohols. The carbonyl functional group can be converted to an alcohol through the use of a Grignard reagent. The carboxyl functional group is weakly acidic. Carboxylic acids are prepared by oxidizing an alcohol or an aldehyde, or by reacting CO_2 with a Grignard reagent. Esters are prepared by reacting a carboxylic acid with an alcohol. Amides are prepared by reacting an amine with an electrophilic group, such as an ester, and are fairly unreactive. Amines are prepared by a nucleophilic substitution reaction, a polar alkyl halide, and either ammonia or another amine.

24.6. The Molecules of Life

There are several types of biologically important organic molecules. The most common type is the carbohydrates, or sugars. Because of their complexity, carbohydrates can undergo a wide variety of biochemical reactions. Monosaccharides are sugars that cannot be broken down into smaller groups. Disaccharides can be broken down into two monosaccharides. And polysaccharides, like starch and cellulose, can be broken down into many monosaccharides. Lipids, or fats, are insoluble in water. Fatty acids are simple lipids composed of a long hydrocarbon chain ending in a carboxylic acid. Saturated fats contain only single bonds and are solid at room temperature. Monounsaturated fats contain one double-bond and are liquid at room temperature. Polyunsaturated fats contain two or more double bonds and are also liquid at room temperature. Proteins are polymers of amino acids, linked together by amide bonds. Of the 20 amino acids, all are chiral except glycine. Nucleic acids are the basic structural units of DNA and RNA. Nucleic acids are linked to a sugar through a glycosidic bond, forming a nucleoside. The addition of a phosphoric acid group produces a nucleotide. Nucleotides link to form a polymeric chain that acts as the backbone for DNA and RNA.

Self-Test

1. How many organic functional groups contain nitrogen?
   A. 2
   B. 3
   C. 4
   D. 5

2. The simplest class of organic molecules is the _____
   A. arene.
   B. alkane.
   C. ether.
   D. alkyl halide.

3. 1-chlorobutane and 2-chloro-2-methylpropane are examples of _____
   A. conformational isomers.
   B. structural isomers.
   C. geometric isomers.
   D. stereoisomers.

4. Which of the following molecules would you expect to be capable of possessing geometric isomers?
   A. 1-bromopropane
   B. Butene
   C. Diethyl ether
   D. Phenol

5. Chiral molecules are optically active _____
   A. always.
   B. under most conditions.
   C. under some conditions.
   D. never.

6. Which of the following is not an intermediate in an organic reaction?
   A. Carbocation
   B. Free radical
   C. Optical isomer
   D. Carbanion

7. A reaction intermediate that is electron deficient and neutral is a _____
   A. carbocation.
   B. free radical.
   C. nucleophile.
   D. carbanion.

8. The opposite of an addition reaction is a(n) _____
    A. substitution reaction.
    B. elimination reaction.
    C. free-radical reaction.
    D. oxidation-reduction reaction.

9. A bulky functional group may cause an organic reaction to be _____
    A. thermodynamically favored.
    B. kinetically unstable.
    C. exothermic.
    D. sterically hindered.

10. Alkanes undergo catalytic cracking, which is an example of a(n) _____ reaction.
    A. Grignard
    B. elimination
    C. pyrolysis
    D. substitution

11. Which of the following is not prepared by the oxidation of an alcohol?
    A. Esters
    B. Carboxylic acids
    C. Aldehydes
    D. Ketones

12. What two types of chemicals are mixed to form an ester?
    A. A carboxylic acid and a base
    B. A carboxylic acid and an alcohol
    C. A base and an alcohol
    D. Two different carboxylic acids

13. Which of the following is not a monosaccharide?
    A. Glucose
    B. Sucrose
    C. Fructose
    D. Galactose

14. Lipids are soluble in water.
    A. True
    B. False

15. The bond between the carbon and nitrogen joining two amino acids is called a(n) _____

    A. protein bond.
    B. chain linkage.
    C. amide bond.
    D. peptide linkage.

Answers: 1. B; 2. B; 3. B; 4. B; 5. A; 6. C; 7. B; 8. B; 9. D; 10. C; 11. A; 12. B; 13. B; 14. B; 15. C